World War II
Aerial Navigation Training
&
The Flyboy Heroes of Selman Field

To Nan—
Do hope you enjoy.

World War II
Aerial Navigation Training
&
The Flyboy Heroes of Selman Field

by

Richard Chardkoff

Richard Chardkoff

Center for Louisiana Studies
University of Louisiana at Lafayette
2008

Photographic Credit:

All of the photographs reprinted in this publication are used courtesy of the Selman Field Historical Association; they were originally donated to that organization by Mr. Walter Kaffka.

Center for Louisiana Studies
University of Louisiana at Lafayette
2008 by Center for Louisiana Studies
P.O. Box 40831
Lafayette, LA 70504-0831

(paper)
ISBN: 1-887366-81-4

Library of Congress Cataloging-in-Publication Data

Chardkoff, Richard.
World War II aerial navigation training & the flyboy heroes of Selman Field / by Richard Chardkoff.
p. cm.
Includes bibliographical references and index.
ISBN 1-887366-81-4 (alk. paper)
1. Selman Field Navigation School (Monroe, La.) 2. Selman Field Navigation School (Monroe, La.)--Biography. 3. World War, 1939-1945--Aerial operations, American. 4. Flight navigators, Military--Training of--United States--History. 5. Flight navigators, Military--United States--Biography. I. Title. II. Title: World War Two aerial navigation training and the flyboy heroes of Selman Field. III. Title: World War 2 aerial navigation training and the flyboy heroes of Selman Field.

D790.2.C53 2008
940.54'4376387--dc22

2008014867

http://cls.louisiana.edu

This book is printed on acid-free paper.

Table of Contents

Acknowledgements

Selman Field, the largest navigation training facility in the United States during World War II, and the role it played are fast fading into history. Those brave young men who passed through its navigation school and set off for war have, for the most part, been forgotten by the present generation.

This book is an attempt to honor the over 15,000 men who trained at Selman, including the 1,480 casualties. Sadly, many of these young men never had the opportunity to marry, have children, or grow old gracefully. Their ultimate sacrifice serves as a lasting legacy to those of us enjoying freedom today.

This book traces the growth of Selman Field from its origins to its final deactivation in 1947. It describes the challenging core curriculum the men struggled through, as well as the demanding physical fitness regimen. An important chapter of the story of Selman Field concerns the warmth and hospitality the small, sleepy, Southern town of Monroe, Louisiana, offered as its citizenry opened their homes and hearts to the young cadets. In conclusion, a few of the personal experiences of some of the men who were stationed at Selman are included. Their narratives, which depict only a microcosm of the experiences of thousands of airmen who graduated from Selman Field, represent the drama, intensity, and heroism of our airmen who served in the European and Pacific theaters of combat.

I would like to express my appreciation to my wife, Joan, and King Trousdale for critically reading and editing the manuscript. Joan, especially, reviewed every draft and offered countless suggestions for improvement. Without her assistance, readiness to help, and dedication, this book would never have been completed. My thanks also go to Jeanette Herren, Bob Ford, Swansey Kotz, and Wilma Rainwater, not only for lending me many of the original Selman Field documents, but also for reviewing the manuscript. Jeanette tirelessly aided me in combing through hundreds of World War II-era photos of Selman Field and in locating documents, old newspaper clippings, and books, which helped enormously to shape the final product.

Prologue

When my family moved to Monroe, Louisiana, in the early 1970s, we soon heard about our new hometown's two important claims to fame: first, Monroe was home to the family whose ancestor, Joe Biedenharn, was the first to bottle Coca-Cola; secondly, Monroe was the place where Delta Airlines began as a small crop-duster outfit.

Not long after our arrival we learned of another important distinction: Monroe had been the location of Selman Field, the foremost school of navigation in the United States during World War II. We were first told about Selman Field by a delightful older couple, Clifford Strauss and his wife Roslyn. They had been kind to us and as a result we invited them to be among the first guests in our new home. During dinner, they started to reminisce about Selman Field and their acquaintance with some of the airmen. They told us they would invite the young cadets fortunate enough to be issued weekend passes to spend the weekend in their home. To make room for their guests, they would crowd all three of their daughters into one bedroom. One of their daughters recalled having to share a small bedroom with her grandmother and two sisters, which meant sharing a single bed with her twin sister. After the young men completed their training and shipped out, the Strausses stayed in touch with them; they would write the men and send them care packages. There was only one catch: there was a military rule that the soldiers had to request an item in writing before it could be mailed to them. Therefore, Mrs. Strauss would write "her boys" asking if they remembered what good chicken soup her neighbor, I.W., made. They wrote back saying they would love some of Mr. I.W.'s chicken soup. Mrs. Strauss prepared the cans herself in the family kitchen, even though her husband owned Strauss and Son, a wholesale grocery and spirits distribution company; however, the cans labeled soup actually contained I.W. Harper bourbon. She then sent the cans to the grateful navigators stationed overseas.

Many years later, I was honored to be asked to write the history of Selman Field. As a history professor, this opportunity proved irresistible. Of course, the story Mrs. Strauss told me came to mind. It made me realize just how much the lives of Monroe residents were touched by all those who had been involved with this airfield.

After word of my project spread by means of the *Selman Field Beacon*, the official newsletter of the Selman Field Historical Association, I was immediately inundated by material. Besides the documents from the Historical Society, I started to receive e-mails, letters, and telephone calls from former navigators all around the country and even beyond, as some of the Free French had also been trained at Selman Field.

Then, in April 2005, I attended the 11th Selman Field Reunion. Airmen returned to Monroe from all parts of the country for the opportunity to reunite with their comrades. I soon realized the urgency of the task. It was said that World War II veterans were dying at a rate of about one thousand a day. At the first reunion in 1985, there had been over five hundred attendees; just twenty years later, there were only thirty. The organizers of the reunion set up interviews for me with the veterans. Everyone had a story to tell.

During the reunion, there was a concert by a group from the Pentecostal Church in Jonesboro, Louisiana. Dressed in World War II-era clothes, the singers sang songs popular during the 1940s. As they sang such songs as "I'll be Seeing You," "America the Beautiful," and "I Remember You," a slide presentation showed period pictures of Selman Field and the men who trained there. "Taps" played as pictures of the Memory Wall, bearing the names of those brave men who had died during the conflict, flashed on the screen. When the performance drew to a close, there was not a dry eye in the house. I promised myself I would do all I could to do justice to the story.

Here then is a brief history of Selman Field and a few of the stories of the over 15,000 men who trained there. Unfortunately, the most compelling stories probably died with the 1,480 men who perished doing their duty. This book is dedicated to their memory. It is my hope that by telling some of their stories, I am honoring all the men who served so valiantly in order to safeguard our freedom.

CHAPTER I

A School Is Born

With the outbreak of World War II imminent, choices had to be made as to how to best protect the national security of the United States. Against a backdrop of limited time and available resources, the creation of a large, well-armed, and well-trained fighting force was essential.

Both the German strategy in the Spanish Civil War and the invasion of Poland and France made clear that airpower would be decisive in the coming combat. Before 1935, the Army Air Corps had been a small nondescript entity under the umbrella of the Signal Corps. There was no specific training program for either bombardiers or navigators. Following the German invasion of France in June 1940, the War Department began a serious study of how to increase the role of Army aviation. A Committee on Training in Specialized Schools was created under the leadership of Capt. Norris B. Harbold. Its mission was to create a curriculum and coursework for the specialized training of navigators, bombardiers, and aerial gunners. The result was a ten-week course consisting of seventy-two hours of flight missions, one hundred hours of training in military protocol, physical exercise, radio codes, and forty-eight hours of gunnery practice.[1]

This plan was first put into action in July 1940, when bombardier and pilot training began at Lowery Field in Denver, Colorado. The original idea was to utilize graduates of this program as instructors for proposed navigation schools at Barksdale Field in Shreveport, Louisiana, and Ellington Field in Houston, Texas. However, there were still not enough planes and, due to conflicting demands, the training of navigators remained a low priority: all available trained personnel had to be assigned to combat units. Therefore, creativity was called for to solve the problem. The Army Air Corps held a series of meetings with representatives of Pan American Airways, which had the largest single group of trained trans-Atlantic navigators in the country. The plan was for Pan American navigators to act as instructors for the Army. Army personnel would serve as assistant navigators on Pan American flights, working under the supervision of the commercial navigators. The Army would provide housing and food service for its trainees. The airmen were to be based in Coral Gables, Florida, a major Pan American hub. The contract required Pan American to offer a twelve-week training course for three classes of fifty students each; the classes would gradually increase to one hundred students. New classes would begin every six weeks. The Army worked closely with Pan American officials to standardize the training program as well as the types of instruments and equipment to be used in instruction.[2]

The first class of forty-six navigators finished training in November 1940. The first military school for navigational instruction had just been established at Barksdale Field, though this school was primarily for pilot training. Several of the new graduates of the Pan American program ordered to Barksdale were to undergo more technical education to prepare them to be future instructors—the teaching nucleus at the navigation school when Harbold took command. Capt. John Egan was appointed assistant director of the fledgling school and Lt. Lawson Moseley was also ordered to join the instructional staff. They revised the course format, compressing training into ten weeks; classroom teaching was expanded to 210 hours, including 100 hours of physical and military training, followed by seventy-two hours of flight instruction.[3]

On June 20, 1941, six months before the attack on Pearl Harbor, the U.S. Army Air Corps was created as a separate command. Nevertheless, it still remained underfunded, lacking both planes and personnel. A short time later, the Corps in turn established the "Flying Training Command," with authority to implement training programs for pilots, bombardiers, and navigators. Barksdale came under the jurisdiction of the Flying Command.

Major problems arose almost immediately. There was a shortage of equipment, especially aircraft that had been configured for navigational instruction. Furthermore, due to a lack of qualified instructors, navigation and pilot training had to be combined.

Following the attack on Pearl Harbor, the pressure for increased military might intensified, and the War Department stepped up its efforts. Mirroring Germany's successful use of air power early in the war, the strategy of the War Department was to crush the enemy's war-making capacity and ability to mass produce weapons. At the same time, the War Department hoped to weaken civilian morale in order to undermine the people's will to carry on the war. The Army envisioned a massive long-range bombing campaign. This required an unprecedented build-up of planes and personnel. The Army Air Corps ultimately envisioned a force of over two million men. To assemble, train, and equip such a massive force would require the creation of scores of new training bases. The military began an immediate search for possible sites for these bases to train the hundreds of thousands of needed technicians, mechanics, pilots, bombardiers, and navigators. Southern sites received priority as their milder winters made possible year-round training. Moreover, rural areas were favored as there would be less danger in case of emergency landings or accidents. Monroe, Louisiana, a city of roughly 29,000, was placed on the long list as a possible site for one of these military bases.

It was Col. Lawson S. Moseley who received the orders to scout potential locations for a navigational training school in the Southeast. Moseley had previously been one of Harbold's navigation instructors at Barksdale. He had since been promoted to lieutenant colonel and posted to Maxwell Field in Montgomery, Alabama. However, because of his time at Barksdale, he was familiar with Monroe. In making his decision, he was instructed to take into consideration the presence of existing airports, the availability of a labor force, ample acreage, community support, favorable flying conditions, as well as the condition of road and rail connections. Several southern sites were considered, including Lafayette, Louisiana and Greenwood, Mississippi. All potential locations had to have good weather, little ground fog, and an absence of snow days.[4]

In late February 1942, Colonel Moseley made his first visit to inspect Monroe. There he met with local political leaders and conducted an evaluation of Selman Field. The field had been named in memory of a Monroe native and Naval Academy graduate, Lt. Augustus J. Selman. Selman had died in 1921 as a result of an airplane crash in Norfolk, Virginia, where he was serving as a naval aviator. The field, six miles east of Monroe, consisted of a single runway surrounded by cotton fields. Although Selman Field lacked infrastructure, it did possess adequate acreage and was near important rail lines and roads.[5]

The airport had been used since the fall of 1939 as a pilot training program for the Army under the recently-passed Civilian Pilot Training Act. An entering class of ten students attending Northeast Junior College of Louisiana State University began pilot training. Instruction was provided by a single civilian instructor using a Piper Cub with a fifty horsepower motor. The program grew quickly; by the fall of 1940, two additional instructors and four more planes had been added to meet the growing demand. By September 1941, the training program enrolled fifty students every semester and the number of planes used for instruction had increased to eight. To accommodate this growth, a flight school hanger was constructed to house the planes.[6] One of the instructors, Earle Barham, a native of nearby Oak Ridge, Louisiana, enjoyed buzzing his family home with low flyovers to alert his family to his presence. His mother later recalled being "frightened to death" by those unexpected visits.[7]

Moseley met with Mayor Harvey Benoit, the president of the Ouachita Parish Police Jury, George Pipes, the parish engineer, Joe Ford, and the president of the Monroe Chamber of Commerce, A.K. Kilpatrick. All expressed their readiness to back the Army's plans and promised full cooperation.[8]

Moseley was sufficiently impressed to draft a favorable report. He also ordered a follow-up site inspection by specialists from Maxwell Field. In early March 1942, Lt. Col. John H. Boushall and Maj. O.M. Baldinger were sent to Monroe. Boushall conducted a detailed survey of housing needs; he concluded that the enormous need for living quarters for hundreds of men could be satisfied with the help of the Monroe Housing Authority, which pledged its full support. Major Baldinger, the field engineer at Maxwell, had the crucial task of conducting a topographical survey of the proposed site, including soil tests and an examination of drainage. This report was also favorable to the Monroe site. Local reporters later credited Baldinger's report as being the decisive factor in favor of Selman over competing training sites at Lafayette, Louisiana, and Greenwood, Mississippi.[9]

On March 25, 1942, the War Department selected Monroe as the site for its new School of Navigation. The airstrip was owned by the Ouachita Parish Police Jury, which had leased it to a crop dusting outfit, Huff Daland Dusters, soon to become Delta Airlines. On June 15, 1942, the War Department designated Selman Field as an active military base. Two days later, on June 17, the federal government lay claim to ownership of the field, some 785 acres, through eminent domain; it paid the police jury $267,500 for the property. An additional 1,857 acres was purchased from forty-one separate owners for $132,491.26.[10]

However, even though the runway could handle small Delta passenger planes and crop dusters, it could not accommodate the larger, heavier military aircraft. It needed to be expanded and heavily reinforced. Construction bids were advertised, and the Kansas City firm of Black and Veatch was awarded the contract of over five million dollars.[11] Now it was time for the real work to begin. The proposed base would be the only all-inclusive navigational training facility in the United States, offering pre-flight and advanced navigation training to all cadets. Upon completion of the program, they would receive their navigator's wings and be commissioned as second lieutenants.

Col. Norris Harbold, a member of the Site Selection Board that had chosen Selman Field, was ordered to Monroe to serve as project manager and activate the base as soon as possible. Harbold, who had been the director of training of the Advanced Navigation School at Turner Field in Albany, Georgia, arrived in early June. He was soon joined by a half dozen officers from the Army Corps of Engineers.

Harbold, then thirty-six years old, was a career army officer. He was from a small town in Nebraska where he had grown up dreaming of a military career. In high school, he had excelled in athletics. After receiving an appointment to West Point, Harbold continued to reap athletic awards, winning varsity letters in lacrosse and football and earning honorable mention as an end on the All-American football team. He graduated from West Point, where he had first shown his leadership potential, serving as cadet captain and vice-president of the class of 1928. A few months after graduation, he began Primary Flying School at Brooks Field, Texas. He next transferred to the Air Corps at Bolling Field in Washington, D.C., where he began navigational training. This proved to be the most formative stage of his military career. Harbold, one of only three army officers to have studied air navigation as a specialized field, used trial and error to develop an experimental program for navigator instruction; this program had its own curriculum, texts, and training equipment. Harbold then served as an instructor at the Navigation School at Langley Field, Virginia, taking with him the pilot program he had developed at Bolling Field. From Langley Field, he was ordered to Rockwell Field, California, to implement the same program. Here he continued to improve instruction. But, after the second of two navigator classes graduated in 1934, the Army abruptly dropped the program. It did not revive it until 1940, when the war seemed certain. In the meantime, Harbold was promoted to lieutenant colonel; he was dispatched to a number of posts: March Field in California, Nichols Field at Manila in the Philippines, Maxwell Field, Alabama, and Barksdale Field in Shreveport, Louisiana. During these years, he served first as a supply officer and later as squadron operations officer. However, he was intent on seeing navigator training reinstated and wanted to get in on the ground floor when it did. His opportunity finally came in October 1940, when he was assigned to the Air Navigation Training School at Barksdale. There he oversaw the classroom training and graduation of the base's first graduation class in February 1941. His success at Barksdale resulted in his transfer to Turner Field, Georgia, as director of training for the Advanced Navigation School. He remained there for ten months until his transfer to Selman Field. Slender with dark hair, Harbold had a calm manner, quick mind, and steely determination. He arrived in Monroe on May 5, 1942, with a solid background in both navigational skills and training techniques.[12]

The job of transforming empty cotton fields into a working military establishment as quickly as possible presented enormous challenges. Harbold was faced with two parallel tasks: construction and securing personnel to serve as instructors. Harbold immediately began working with a skeleton force of army engineers as well as Black and Veatch civilian contractors. In meetings that lasted from dawn to late evening, they conducted performance surveys to locate streets and buildings, runways, and gas, water, sewage, and electrical lines. But before construction could begin, the field had to be cleared. All civilian planes were moved to a dirt runway several miles away, just north of Black Bayou. The crop dusters were relocated to a temporary dirt runway in Bernstein Park in south Monroe.

Capt. Otis P. Johnson, Jr., chief army engineer, held the task of working with the civilian engineers and architects to keep the construction projects on track within a rigid time frame. After initial surveys were completed, blueprints firmed up, and an inclusive master plan agreed upon, over 2,800 workers were brought in to begin work. Johnson closely monitored every project and gave final clearance for all building designs, specifications, and estimates. Days were long and arduous under the broiling Louisiana sun. Two construction companies, T.L. James Company of Ruston, Louisiana, about thirty-five miles away, and Forcum-James of Dyersburg, Tennessee, were subcontracted to build the post. T.L. James was to construct the buildings and do most of the work on the runways and Forcum-James was responsible for the infrastructure; this included underground conduits for gas and water distribution, irrigation ditches, and the sewage system.[13]

The first priority was the construction of a new concrete runway. The existing runway could not be used since a river levee would block its extension. New roads and irrigation ditches began to spread out across the cotton fields; rail spurs to carry building materials to the site were built, and work was begun on the infrastructure and utility system. In addition, construction began on a storm water drainage system, including a levee around the field, two pump stations, and a sewage disposal plant.[14]

At the outset, serious logistical problems arose. First, the fields did not drain well after heavy rains. The subsequent soil tests, which were necessary before the runways could be laid and graded, indicated an almost insurmountable problem. Before the engineers could lay the concrete runways, the soil had to be highly compacted. However, when they attempted to do this with huge rollers, the subsurface water combined with the surface dirt to form a spongy mix incapable of supporting any weight. With time running out, options were limited. A different location for the runways seemed out of the question. Nor was it practical to pump out the immense amount of subsurface water, which would have resulted in huge sinkholes. The engineers realized they would have to come up with an unorthodox solution. After much deliberation among the engineers, a radical idea emerged. Harbold later described the concept:

> The only possible solution was to float the runways. This was accomplished by grading the runways to the level required for the concrete slab. The thickness of this slab had been increased by several inches over the original specification. Once the runways were graded, they were allowed to crust and the agitated water was allowed to resettle at its original level.[15]

The runways at Selman Field still float on a field in Ouachita Parish to this day. The soil remains dry only because of a slant levee and an elaborate drainage system.

Work continued throughout the summer of 1942. The base was officially activated on June 15, 1942, but the field's strength consisted of only Harbold, Captain Johnson, and four enlisted men. Their priority was the construction of the runways. The north-south runway was completed by August 31, 1942, while the east-west runway was finished by December 1, 1943. Eventually there were four runways, each 5,000 feet long and 150 feet wide, with connecting taxiways and 300,000 square miles of apron, all thirteen inches thick.

Numerous structures in various stages of completion began to dot the field. The first building to be completed was the fire station; with the arrival of two brand new fire trucks, fire protection was in place.[16] Unseasonably heavy rains in June and July forced T.L. James Construction to resort to other rather unconventional means in order to maintain the frantic pace of construction. To reach the most inaccessible parts of the swampy sea of mud, the workers brought in horses and mules. Tom Lofton, one of the civilian construction superintendents, rode a pony named Spot. Spot was taken to and from the base by truck and stabled in a warehouse on DeSiard Street. Lofton figured that the cost of maintaining the pony was actually less than for a vehicle. In addition, four army mules were put to work hauling a sled containing oil and gasoline drums. The fuel was needed for the bulldozers and other heavy equipment. The four mules were worked in teams of three, with each mule given a day off every fourth day. They were pastured on the field that provided most of their feed. T.L. James figured that, at about one dollar a day, the mules were very cost effective.[17]

The taxiways and aprons on the far eastern periphery of the base were still incomplete and there was still no lighting. In spite of these circumstances, the urgency of the war effort resulted in the use of the not yet completed north-south runway.

During the first two weeks of August, everything finally seemed to come together. First, the population of the base increased. On August 1, fifteen enlisted men from Maxwell Field in Alabama reported for duty. The following day, they were joined by an additional twenty-five men of the Medical Corps detachment, also from Maxwell. But while the field was filled with activity during the day, by night it was completely deserted. Since the men still could not live or eat on the post, arrangements were made with the Monroe City School Board to house them in temporary accommodations in locker rooms under the Neville High School football stadium.[18]

The men in charge of finding off-base housing for the new arrivals were Lt. D.H. O'Connell, a former classmate of Colonel Harbold's at West Point, and Sgt. Bradford Blue, who had served under Harbold at Maxwell Field. They also arranged for the men to have meals with local families or eat at restaurants and ride with civilians to and from the base. Finally, on August 8, this problem was resolved when Mess No. 1, although not quite complete, was pressed into service. First Lt. William Kelly, the mess officer, and his technical sergeant, Prince Peevy, served the first meal. They did their best to make the meal fit the special occasion: baked ham, boiled potatoes, Harvard beets, and green peas. Peevy, a graduate of the Army Cooks' and Bakers' School at Fort Benning, Georgia, had served in the Army for two years. Although he worked hard and tried to vary the menu, many of the cadets still found the food less than appealing. Moreover, there was never enough silverware to go around.[19]

The same day that the first meals were served at Selman, the first barracks were also ready for use. The troops temporarily quartered at Neville High School, were now relocated to base. The one-story open-bay wooden barracks could house forty men. They were joined by seven officers and fourteen key non-commissioned officers from Maxwell Field, who had been part of the advanced cadre at the pre-flight school.

Conditions were primitive; the barracks had wooden frames with tarpaper siding and no insulation, plumbing, or ventilation. In the winter months, warmth came from a coal-burning stove or gas space heater. As the barracks had not yet been wired for electricity, Coleman lanterns provided the only illumination. The interior was bare except for a stack of canvas cots and some bedding. There was not even mosquito netting for the windows. Latrines were placed centrally every four to five barracks apart.[20]

The small advance team arrived with orders to prepare for the transfer of hundreds of pre-flight navigation students and advanced navigation cadets from Maxwell Field to Selman. The team was given less than a week to make certain base housing and classroom materials were ready for the new arrivals. Their intention was to accomplish the move so efficiently that not a single day of classroom instruction would be lost.

On August 11, following a two-day journey from Turner Field in Albany, Georgia, a military convoy of fifty vehicles arrived. On the convoy were five officers and 166 enlisted men. Forty-six enlisted men also arrived by private car later the same day. Then another twenty-one enlisted men of the Quartermaster Corps arrived the following day. Soon military specialists began arriving from all over the country; they were there to assess the needs for basic services the base would require with the huge influx of expected personnel.[21]

The final phase of the plan was then activated over the weekend. In a one-day operation, a twenty-three car troop train brought 728 cadets, including enlisted men and officers, from Maxwell Field, Alabama, a distance of 400 miles. There was also a twenty-five member medical detachment and additional enlisted men from Maxwell.[22]

The move was not without problems. The troops were provided with only a box lunch and crowded into railway coaches that were stiflingly hot. What was originally scheduled to be a twelve-and-a-half-hour trip turned into a twenty-two-hour ordeal. The cadets arrived tired, hungry, and drenched in sweat. They were covered in soot and ash that had blown into the rail cars through the open windows from the coal-fired steam engine. After being ushered off the train, the men were marched, baggage in hand, down a dusty unpaved road from the train to the mess hall. Following a hurried meal, they were marched to temporary barracks and told to set up their canvas cots. Lister bags, containing twenty gallons of drinking water trucked in from Monroe, were hung from tripods outside the barracks. The toilets and the showers worked, but there was only cold water. Furthermore, the water was oily as most of the pipes had been recycled from oil and gas fields. When the men showered, they were left covered with a greasy slime that they had to wipe off with their towels. It was not unitl October that two water tanks were erected to supply clean chlorinated water to the Post.[23]

Later that evening, August 15, ninety-eight additional military and civilian instructors, administrative officers, cadets, and enlisted men arrived from Maxwell by private car. The next day five hundred desks and chairs were moved into the temporary buildings to prepare for classes the following morning. At the same time, Harbold's administrative staff moved into what had been the Delta Airlines passenger terminal; it had been hastily refurbished with a minimum of amenities. The office space was converted into a kitchen and dining room with a table and four chairs. The waiting room became a bedroom furnished with hospital beds; clothes racks served in place of closets.[24]

The transfer to Selman Field was completed the following Monday night when cadets from the Advanced Navigation School at Turner Field in Albany, Georgia, arrived by train. Gas masks and luggage in hand, they were immediately marched in columns of four to the mess hall for a late supper. These cadets, however, had found the trip less taxing than the previous group; in addition to the box lunch provided by the army, the train had stopped in Vicksburg, Mississippi, where the Red Cross had greeted the troops with fruit juice, sandwiches, doughnuts, and cigarettes.[25]

Even though construction was not yet complete, Harbold nevertheless thought that it was time to begin operations upon the arrival of the cadets. Accordingly, he had seventy-three twin-engine Beechcraft AT-7s, specially modified for navigational training, flown in from Turner Field, Georgia. They were accompanied by a contingent of maintenance support personnel as well as a number of pilots and other enlisted men and officers.[26] Although the north-south runway was not completed until August 31, Colonel Harbold began accelerated pre-flight training at the navigation school on August 19, the day after the cadets arrived from Maxwell Field. The first class totaled 246, including 239 cadets and seven student officers. Instruction took place in makeshift classrooms; yet both the course of study and physical training programs were rigorous.

On September 22, navigational training flights finally began. The drone of planes could be heard all over the base at all hours of the day and night. From August to December 1942, permanent construction gradually replaced the hastily-erected temporary structures. In October, the cadets were moved to the just-completed barracks located near the classrooms. The new barracks each housed twenty-five to thirty cadets. The buildings were partitioned into three or four rooms, with six cadets assigned to each section. There was electricity as well as oil heat, but the cold showers and the outdoor latrines remained a constant source of irritation. Eventually, there were 623 structures on the base in various stages of completion.[27] In November 1942, a recreation hall, post exchange, and station hospital were ready for service; however, although the building was completed, the hospital still had no beds and the small medical staff had to rely on assistance from the Ouachita Parish Medical and Dental Societies. Indeed, the first Army nurse did not report for duty until January 1943. A mess hall for newly-arrived black troops opened in time to serve the 1942 Thanksgiving dinner. The officers' club and attached officers' mess opened on

December 1. Finally, a post exchange for the black enlisted men opened by mid-December.[28]

Nor were the spiritual needs of the troops neglected. Confirmations, memorial services, and weddings all were conducted on the base. The north chapel for Protestant religious worship was completed in November, soon followed by the south chapel for Roman Catholic and Jewish worship services. Both chapels, identical to one another, faced each other. The first wedding held on the post was in a temporary post chapel on August 26, 1942, just weeks after the base was activated. Local churches worked closely with the post chapels in welcoming the cadets and meeting their spiritual needs.[29]

Once the basic post structures were in place, the next step was to provide amenities for the men during their brief periods of downtime. A post theater was ready for pre-flight cadets in time for a 1942 Christmas dinner show. In 1943, a post library was placed temporarily in the guest reception house; later, the holdings of 3,000 books were permanently relocated to the Service Club.[30]

Soon after the creation of the base, two lieutenants, John Legrand and Louis Ravielli, while working on a navigation chart, devised a symbol for the new base. Legrand came up with the idea of a mascot to symbolize the constant "groping around" of the cadets as they attempted to navigate with the aid of the crude instruments then available. After some thought, he came up with the name of "Joe Grope." Ravielli then developed and drew a camel-faced character for the symbol and thereafter "Joe Grope" appeared on all Selman Field printed materials and insignias.[31]

A post newspaper and radio show provided both base information and world news to the cadets. The radio show "Selman Field – On the Beam," produced with the assistance of a local radio station, KMLB, was managed by the former director of the Monroe Little Theater. The newspaper *True Drift*, published by the *Monroe News-Star*, covered base happenings.[32] Other extra-curricular activities included a small arms firing range, a track and field area, an outdoor boxing ring, a skeet and archery shooting range, a post gymnasium, a 400-yard, twenty-one-barrier obstacle course, and swimming pools for black troops and officers as well as for white troops and officers. In addition, there was a new officers' club. Next to the post exchange there was even a four-lane bowling alley built in the shape of a blimp hangar. Free cakes, coffee, doughnuts, and finger food were provided by local women volunteers who staffed a new base canteen under the auspices of the Red Cross. The post also provided boats for those who wished to enjoy the natural beauty of nearby Bayou DeSiard. In September 1942, the first WAC, Joan Dickson of Houma, Louisiana, arrived. She was the first of more than a score of WACs transferred from Turner Field, Georgia. Following their arrival, a beauty parlor opened for their use.[33]

The base also benefited from its location in Monroe and the goodwill of its citizens. Local organizations and townspeople donated materials and labor to help with a number of post projects. Funding for materials came from a $2.00 contribution from officers and $.50 from cadets and enlisted men, while local businesses donated their time. For example, the Stovall Drilling Company excavated the swimming pools. Unfortunately, though, while drilling the well for the pools' water supply, the drilling rig penetrated a salt dome and the water came out salty.[34] Local garden clubs planted shrubbery and grass, making the base more attractive and also controlling the mud and dust. Townspeople contributed more than 2,000 plants and fifty acres of grass to the base. It took the nurseries in Monroe over 300 truckloads to deliver the plants and shrubs that were planted by the airmen over several months.

In another effort to reduce the amount of dust, the roads were sprayed with tar and coated with pea gravel. In addition, wooden sidewalks were built to connect the major buildings. Unfortunately, many of the planks tended to float away following heavy rains.[35]

Although the cadets had limited free time due to the rigorous academic load, the post still sponsored a wide variety of intramural and off-base athletic competitions. Activities included boxing, football, softball, basketball, track, tennis, and volleyball for both black and white troops. Games and matches were scheduled regularly with nearby military posts and area colleges, especially the Northeast Junior College in Monroe. Whenever possible, the public was invited to attend the sporting events. The base football team, the Cyclones, attracted large crowds. In time, 3,500 to 4,500 fans were on hand to cheer the team.[36]

The most popular site for on-base entertainment was the theater. There was a continual slate of dances, plays, concerts, both pop and classical, all of which helped boost morale. Entertainment was provided both by base personnel, many of whom had been professional entertainers or musicians before the war, as well as by community groups and touring USO shows. USO tour groups made monthly stops to perform at Selman. Beginning in 1943, the newly organized post band presented weekly concerts every Friday night at the service clubs; the concerts were followed by a dance. Area civic and service clubs also joined in to provide entertainment. The Shreveport-Bossier Kiwanis Club presented a two-hour musical revue, the minstrel show "Flying High," which received a standing ovation. Students from Centenary College in Shreveport, community groups from the nearby towns of Bastrop, Winnfield, and West Monroe, as well as towns as distant as Baton Rouge, all presented plays and musicals.[37]

Another popular show was "Steppin High," an all-black USO road show that earned rave reviews and played before a packed house of over 2,000. Among the visiting celebrities was Bob Hope, who arrived in April 1943, to perform at Selman Field, including a live broadcast of his "NBC Pepsodent Hour" radio show. He was accompanied by Francis Langford, Jerry Colonna, Vera Vague, and Skinney Ennis and his orchestra. Before leaving Monroe, Hope also held a successful war bond rally at Neville High School. All those who wished to see Hope were required to purchase a war bond of any denomination. Hope is credited with selling over $1,000,000 in bonds in support of the war effort. Other celebrities who performed or held war bond rallies included the Glen Miller Band, the Tommy Dorsey Orchestra, Ozzie and Harriet, Lt. Hank Greenberg, former Detroit Tiger's batting ace and first baseman, and motion picture star Errol Flynn.[38] Still other entertainers came for personal reasons. Shelley Winters, at the time a bit actress just starting her film career, visited her husband, Mack Mayer, who was stationed at Selman. Both Mayer and his wife were to become well-known film stars in the post-war years.[39] The airmen eagerly anticipated the performers and several shows were forced to limit admission due to the huge demand.

There were also some off-base locations that were popular with the troops. A country western bar, built in the 1930s and known locally as the 3 Mile Inn because it was located three miles from the five points intersection on DeSiard Street, was converted into a non-commissioned officers' club. It was only a few miles from the base and although the road between the two was gravel and filled with potholes, there was a steady stream of traffic to and from the base. However, the building did have one peculiar feature: the dance floor extended out over the bayou and had a glass floor. As a result, whenever there were dances, military police had to be posted to keep local teenagers from swimming across the bayou to peer through the floor and up the girls' dresses.[40] Today the former club is a popular local restaurant known as the Waterfront Grill; though remodeled, it has retained the charm of its prewar architecture.

A number of popular off-base venues were located downtown in Monroe. The Pre-Flight Club in the Cameo Room of the Virginia Hotel opened on August 21, 1942. It was used for dances on the roof garden. Close by, the Kickapoo Lounge in the Monroe Hotel, and the Francis Hotel bar were popular watering holes for the airmen. Nearby, a USO was established in the H. Mickel Building on Catalpa Street. Red Skelton and other Hollywood personalities performed there for fund-raisers and war bond rallies. Also close by, a former Jitney Jungle grocery store was converted into a USO for black troops.[41]

Other local establishments were popular with the troops as well. However, after fights broke out with local boys and prostitution grew rampant, lists were posted placing the Spanish Inn, the Columbia Hotel, the Golden Gate Bar, the Brass Rail, and the Green Derby off-limits to military personnel.[42] The lists were constantly updated and revised as businesses complied with military requirements. As a result, dances and buffet suppers were held on base fairly regularly to take up the slack.

Hundreds of local girls attended the receptions and dances, either on-base or at clubs in town hosted by the military. The first of many social gatherings occurred on November 24, 1942, when the post opened for civilian inspection. Over 500 local girls showed up at a buffet supper and dance hosted by 1,000 enlisted men. The mess prepared over 700 pounds of turkey, 1,100 pounds of ham, 204 mince and apple pies, and 120 gallons of coffee. The post also provided entertainment: a pre-flight wing review, an exhibition flight, and a precision drill all added to the excitement.[43] Thereafter, on many occasions, buses were sent to the USO to pick up girls to take them to Selman Field for dances. For the local girls, a new social horizon opened up and many relationships formed, some of which resulted in marriage.

The citizens of Monroe opened their homes and hearts to the young cadets, many of whom had never been in the South before. They offered hospitality to the cadets throughout the war years. George Armstrong, a twenty-three-year-old pilot from Washington, D.C., just graduated from the Army Air Forces Advanced Flying School in Columbus, Mississippi, on February 16, 1943. The following day he was one of thirty newly-commissioned second lieutenant pilots who arrived in Monroe to serve as pilots to fly the navigator cadets. After a long, hot ride aboard a Greyhound bus, they pulled up about 6:00 P.M. in front of the old Monroe Hotel and Coffee Shop downtown on South Grand, where the tired airmen were pleasantly surprised by the warm greeting, and sat down to a delicious meal at reasonable prices. Their first impression of Monroe's hospitality was reinforced by many incidents of kindness during their stay at Selman. Armstrong later recalled that whenever he or one of his fellow pilots went to town in uniform and ordered a cup of coffee, they almost always found when they left the restaurant that someone had already paid for the coffee or that it was on the house. He also remembered that a local gasoline station on DeSiard Street run by two brothers, Joe and Ed Cascio, always managed to keep his old Chrysler in spare tires in spite of strict wartime rationing. While stationed at

Selman, Armstrong met and married his wife, Mattie Glass, who was employed at the downtown Sears.[44]

On weekends, for those cadets lucky enough to get passes to leave the post, townspeople were eager to entertain the troops in their homes. Often these invitations involved some sacrifice, as everything was rationed and sharing meals with the men meant less for the families. Nevertheless, the citizens of Monroe gave generously of both their time and money. The cadets reciprocated by bringing cokes, gum, or candy bars from the commissary. These treats were especially appreciated by the children as these items were rationed and almost impossible to purchase. On holidays such as Christmas, large numbers of Selman cadets were invited into homes in Monroe, West Monroe, Rayville, Bastrop, and other area towns. On Christmas Day 1943, almost 1,000 Selman troops were guests in local homes under a Red Cross program. Families took the young men quail and deer hunting or fishing and boating on the area's lakes and bayous. For cadets from large Northern cities, this was a brand new experience. Cadets were even married in local homes with the homeowner sometimes serving as best man. Community members were often given the honor of pinning wings on the cadets after they graduated when no immediate family was present.[45] The townspeople wanted to do whatever they could to support the war effort and they felt that by supporting the troops who were stationed in their town that they were doing everything they could to help. After the cadets left for war, they often stayed in contact with their hosts. When these young men were killed, their loss was mourned as if they were part of the family.

With Army Air Force demands for trained navigators critical and the base active, Harbold, a veteran instructor, stressed and accelerated the rigorous military training. Under his supervision, the base grew into a full-fledged military installation. By December 1942, base personnel climbed to 6,179. However, Harbold's time at Selman was limited by the demands of war. After less than a year at his post, in January 1943, Harbold received orders to report to the training command at Fort Worth, Texas. He was named the new assistant to the assistant chief of staff for operations.

Harbold's replacement, Col. Earl L. Naiden, arrived almost immediately. Naiden, a forty-nine-year-old native of Woodward, Iowa, was likewise an aviation pioneer. Like Harbold, he was a graduate of the U.S. Military Academy at West Point. Shortly after his graduation, he served under Gen. John "Blackjack" Pershing in the Mexican border campaign, in search of the elusive Pancho Villa. After the United States entered World War I, Naiden joined the Army Signal Corps, the rather small and neglected branch where military aviation was clustered. It was during this time that he became aware of the full potential of aviation to military strategy. He saw the future and wanted to be part of it. After service in France and Italy, Naiden studied and taught aviation at elite military schools in the United States and Europe, gradually working his way up the military chain of command. Following the attack at Pearl Harbor, Naiden was appointed chief of staff of the Army Air Force in the India-China-Burma theater. He had the difficult task of organizing a supply route from India to China to supply the Nationalist Chinese forces. At the same time, with few available resources, he was to command all tactical air operations against the Japanese advances. He spent eight months organizing the India-China Ferry Command, considered crucial following the Japanese conquest of southern China and the cutting off of the Burma Road. With this situation stabilized by October 1942, Naiden was assigned to Craig Field in Selma, Alabama, as the new commanding officer. He had hardly settled in, when in January 1943, he was ordered to Selman Field.[46]

Naiden continued the successful program of instruction begun by Harbold, at the same time making a number of modest changes in the curriculum. Random hazing of the new cadets by officers and upperclassmen was abolished. Moreover, no matter how demanding his duties, he made every effort to get out in the community; he addressed Monroe civic and social clubs to cement the town-base relationship at every opportunity. Naiden had, however, a low tolerance level and he could be abrasive and short tempered. Indeed, at one point in his career after he left Selman, he responded rudely to a British general and was demoted to colonel as a result. As the war progressed, he did manage to earn back his general's stars. But Naiden barely had an opportunity to leave his mark at Selman Field. In August 1943, after only a few months in charge, he was ordered to take over command of the Tenth Air Force. In September 1944, Naiden's plane developed engine problems and crashed near Medford, Oregon. There were no survivors.

Naiden's replacement, Col. John W. Egan, was yet another aviation pioneer. Before taking over command of Selman Field, he had over two decades of study and flight training. Egan had participated in early Army Air Force exercises testing the capability of aerial attacks on naval ships. These exercises had underscored the need for precise navigation training to determine the exact location of bombing targets. Then, when the first Navigation Training School was established at Barksdale Field in 1941, Egan was appointed assistant director. Later he was assigned to Mather Field as director of navigation training. In order to increase his knowledge of cutting-edge navigational training techniques and equipment, he visited RAF navigation training bases in Canada and Britain. After Pearl Harbor, he helped to rewrite the curriculum for navigational training based on the combat-tested British techniques he had investigated earlier. At the same time, he was given the job of helping to select possible sites for all-inclusive navigational training bases. Even though he had never seen it, Selman was one of the sites he had proposed. Thus, ironically, when Colonel Egan arrived at Selman Field in August 1943, to take over base command, he had already played a key role behind-the-scenes in its establishment. A man of quiet self-confidence, Egan was wise, tough, and well-liked by base personnel. He served as base commander of the field until April 1944, and in a dual capacity as training commander until August 1944, when the base was operating at peak capacity. During Egan's brief tenure, base personnel topped 11,000 and navigational education was marked by superior training. When Egan left to become air inspector at Ellington Field in Houston, Texas, navigational training was winding down. Colonel Egan later recalled his time at Selman: "It was a long and interesting flight for one lone eagle. But for me and the rest of us who lived it, it was the only place, and the only time."[47] Following Egan's departure, the last year of the war saw four different commanding officers administer Selman Field, as the mission of the base continued to undergo changes.

In 1942, when navigational training began under Harbold, there was an initial student body of 707 cadets in the pre-flight and advanced navigation training. From August 1942, until October 1945, 15,349 cadets passed through the pre-flight and advanced navigation schools. Of these, 1,870 cadets, or approximately nine percent, died in training accidents or were killed in action in virtually every theater of combat in World War II. A total of fifty-four classes graduated from Selman Field.[48] In the first class, eighty-five cadets completed course requirements, graduating in October 1942. The number of navigators needed was linked to the number of available pilots. There was an increasing need for navigators until mid-1944, when the number of navigators finally reached parity with that of pilots. After that time, navigator training fell off. By the end of the war, over one-third of all navigators flying heavy bombers had been trained at the Monroe training facility.

Even though instructional standards were high, training was not without risk. The achievements of the highly qualified graduates were earned at a very high cost. Nevertheless, considering the hundreds of thousands of hours flown in training exercises both day and night, the number of deadly accidents could have been worse. During routine training, Selman witnessed over 100 major accidents resulting in the deaths of over eighty cadets. More than half of these accidents were attributed to pilot error.[49]

The first fatal crash at the base occurred in October 1942, roughly twenty-five miles north of Monroe; two men were killed in the crash of their PT-19.[50] This cast a pall over the entire post and brought home to the cadets the reality of the danger they faced. April 1943, was an especially deadly month. Eight planes were involved in serious accidents; in one crash, that of a C-60 at Evansville, Indiana, ten men were killed, including three men from Monroe. Their loss was mourned both on the post and in the city.[51] Two months later, in June 1943, five Selman cadets were killed in a mid-air explosion of their AT-7 Beechcraft near Hammond, Louisiana. By August 1943, student navigators were flying over 17,000 hours a month during training flights, thereby increasing the risk of accidents. That same month there were eight major accidents involving eight planes. The most serious was the crash of an AT-7 trainer six miles north of Montgomery, Texas, which killed all five men aboard. Then two men were killed in a crash during take-off of their AT-7 at Moody Field, Georgia, and another student drowned in the crash of his plane fifteen miles below Vicksburg, Mississippi. Only two months later, in October 1943, the number of training hours flown per month surpassed 19,000 hours. That month thirteen planes were involved in ten separate accidents.

This terrible trend continued through November and December. Training hours by then had increased to more than 41,000 a month. During this time, there were nineteen major accidents involving twenty planes. January 1944 was another grim month. Five planes were involved in separate accidents. In the most serious crash, two planes from the base collided in mid-air and exploded upon impact, scattering debris over a large forested area about thirty miles north of Monroe. All ten men aboard the planes perished. In March 1944, nine plane crashes resulted in the deaths of fourteen airmen. The last and most serious single accident occurred on June 26, 1945, near Tampa, Florida, when thirteen airmen were killed in a crash.[52]

In January 1945, the base began downsizing. The pre-flight school was shut down, with the remaining students transferred to gunnery school in Harlingen, Texas. A short-lived program, the Bombardier Redeployment Training Program, was introduced to retrain bombardiers who had served at least one combat tour. The program, lasting twenty-four weeks, was to insure the men retained bombardier skills while introducing them to navigation training. The end of the war in Europe coupled with the surplus of available trained navigators and bombardiers, lead to the termination of the program in July 1945. At the same time, three additional programs were implemented. A new training program known as the Combat Crew Continuation Program began. The program was an attempt to keep the navigators' skills at peak efficiency. Simultaneously, a thirteen-week instructional course geared towards training bombardiers to be navigators on B-26 bombers was put into place. Finally, in August, a tactical training program was started. All the programs were terminated with the surrender of Japan. After that, due to the drastic reduction in training flights, the accidents finally ceased. The summer of 1945 saw three different commanders take charge of Selman Field in rapid succession before being relieved and transferred.

In September, following the end of the war, the Navigation Training School was moved to Ellington Field in Houston, Texas. Only a skeletal administrative staff remained. During the last few months of the war, from May to September, due to personnel shortages, fifty German prisoners of war from the prison camp in nearby Ruston were trucked to Selman each day to perform routine maintenance and mess hall duties on the base.

In October 1945, Selman Field was designated as one of five temporary staging areas to begin processing men, replacing combat troops overseas with troops who would be on duty during the occupation. Three months later, in January 1946, Selman Field was closed in spite of strong protests from Monroe Congressman Otto Passman and from other community leaders.[53] Although Selman reopened briefly as an auxiliary base to Barksdale Field in Shreveport, it was finally and officially deactivated in 1947; it was at that time that the United States Air Force came into being. In 1949, the ownership of Selman Field was officially transferred to the city of Monroe; a recapture clause was included in the document recognizing the transfer that permitted the federal government to use the base in the event of a national emergency. Thus the role of Selman Field and its proud contribution to the allied war victory came to an end.

Vestiges of the post remained even after the base closure. The officers' club and the officers' swimming pool, operated by the Monroe Optimist Club, were open to the public for several years. The post theater, scene of so much popular entertainment, was converted into a community theater and used by the Monroe Little Theater until the early 1960s when a permanent theater was constructed in Monroe.[54] Several buildings were used as temporary classroom facilities by Northeast Louisiana University until the early 1970s. The remaining structures, except for one used as a military museum and another as a sheltered workshop, were demolished. The two runways and two refurbished hangers are virtually all that remain of the once thriving military post.

Endnotes

[1] Norris B. Harbold, *The Log of Air Navigation* (San Antonio, TX, 1970), 78.

[2] History of the 80th Flying Training Wing, USAF Historical Division Archives, Maxwell Field, Montgomery, AL.

[3] Harbold, *The Log of Air Navigation*, 85; Monte D. Wright, *Most Probable Position: A History of Aerial Navigation to 1941* (Lawrence, KS, 1972), 190-193.

[4] *Monroe Morning World*, March 4, 1942, 1.

[5] Robert Hallack, *Selman Field: An Army Air Forces Navigation School* (1984), Special Collections, University of Louisiana at Monroe, 3.

[6] Billie Joiner and Jean Berry, "The History of Aviation at Northeast Louisiana University," unpublished paper (July, 1985), 1-5.

[7] Interview with Senator Robert Barham, Monroe, LA, September 28, 2006.

[8] *Monroe Morning World*, March 12, 1942, 1.

[9] Ibid., April 12, 1942, 2.

[10] *U.S. vs. Ouachita Parish Police Jury, et al*, June 17, 1942, Conveyance Records of the Ouachita Parish Clerk of Court, Monroe, LA, Conveyance Book 297, 11; *Monroe Morning World*, August 28, 1944, 1.

[11] Halleck, *Selman Field*, 4.

[12] *Monroe Morning World*, August 28, 1942, 1.

[13] Ibid., October 11, 1942, 7.

[14] Ibid., October 4, 1942, 7.

[15] Ibid., September 27, 1942, 6.

[16] *Selman Field: World War II: 1942-1946* (Paducah, KY, 1998).

[17] *Monroe Morning World*, July 26, 1942, 6.

[18] Ibid., June 14, 1943, 1.

[19] Ibid., September 16, 1942, 10; interview with Harry Peterson, Monroe, LA, April 29, 2005.

[20] Selman Field Base Operations, "History of Base Operations and Affiliated Departments," July 1942-February 1945, 1.

[21] *Selman Field*, 15.

[22] Selman Field Daily Navigation Log, August 1942, personal collection of Roy Gates, Monroe, LA.

[23] Ibid.; Hallack, *Selman Field*, 6.

[24] *Selman Field*, 19; Charles D. Baylis, ed., *Yearbook of Selman Field, Monroe, Louisiana, 1943*

(Baton Rouge, 1943), 34.

[25] *Monroe Morning World,* September 16, 1942, 10.

[26] Selman Field Base Operations, 1.

[27] Interview with Roy Gates, Monroe, LA, October 11, 2005; Selman Field Daily Navigation Log.

[28] Selman Field Daily Navigation Log; *Monroe Morning World,* August 9, September 6, 19, 1942.

[29] Selman Field Daily Navigation Log.

[30] Ibid.

[31] Selman Field Reunion Brochure, 1993.

[32] Hallack, *Selman Field,* 9.

[33] *Monroe Morning World,* September 13, 1942, 5.

[34] Ibid., June 27, 1945, 1.

[35] Ibid., July 16, 1942; Selman Field Daily Navigation Log.

[36] Ibid.

[37] Ibid.

[38] Ibid., November 3, 1942, 3.

[39] Hallack, *Selman Field,* 11.

[40] Interview with Don Weems, Monroe, LA, September 23, 2004.

[41] Selman Field Daily Navigation Log.

[42] *Selman Field Beacon*, 2:2 (Fall, 1992), 1.

[43] Selman Field Daily Navigation Log.

[44] Interview with Colonel George Armstrong, Monroe, LA, February 23, 2007.

[45] Selman Field Daily Navigation Log. Interviews with Leslie Leibovitz, Rita Blumenthal, Rose Marie Album, and Jean Marcovitz, Monroe, LA, April 30-September 22, 2006. Interviews with Leslie Leibovitz, Rita Blumenthal, Rose Marie Album, and Jean Marcovitz, Monroe, LA, September 17-22, 2004. Interviews with Georgette Bowie and R. Burrell Bailey, Monroe, LA, April 30, 2005.

Leslie Leibowitz met her husband, Irving, in May 1943, when he was a cadet at Selman Field. He graduated in October 1943, as a second lieutenant and was assigned permanently to Selman as a navigation instructor. They married in 1944.

Francis and Abraham Blumenthal, owners of Fields, an upscale women's shop in Monroe, had cadets at their home for dinner every week and took several of them quail hunting and fishing, a first for those from large northern cities who had never had the opportunity to engage in outdoor sporting activities in rural areas.

Jake and Rose Pickens, the parents of Rose Marie Album and Jean Marcovitz, had recently moved to Monroe from Oklahoma where they opened up a men's store. Jake secured a contract to provide graduation uniforms for the cadets and both he and his wife immersed themselves in volunteer activities at the field. Rose ran the Friday night NCO Bingo games and Jake and Rose entertained troops at their home. Indeed, one of their houseguests, Cadet Milton Popkin, was married in their home and Jake served as best man.

R. Burrell Bailey was born and raised in New York. He had never been to the South before being inducted into the Army. After passing through the Nashville classification center, he was ordered to Selman Field in 1943, where he met and married a student nurse at Saint Francis Hospital in Monroe.

Georgette Bowie met Jack Gordon Andrews from Rochester, NY, shortly after his arrival at Selman in 1945 as a cadet. It was love at first sight. Jack proposed on their first date at Walgreen's Drug Store in downtown Monroe. Both were nineteen. She accepted, but when he approached her mother for permission to marry she asked them to wait until they knew each other better and were older. They refused to take no for an answer and drove to Mississippi where they were married.

[46] Richard E. Morse, *Personal Papers in the United States Air Force Historical Research Center: Personal Papers of Earl Naiden, 1911-1944.* 5th ed. Maxwell Air Force Base, AL: US Air Force Historical Research Center,1990.

[47] *Selman Field Beacon*, 2:2 (Fall, 1992), 1.

[48] Selman Field Daily Navigation Log (figures compiled from entire log).

[49] Hallack, *Selman Field,* 8.

[50] Selman Field Daily Navigation Log.

[51] Ibid.

[52] Ibid. (figures compiled from entire log).

[53] Randolph Jones, "Otto E. Passman and Foreign Aid" (M.A. thesis, University of Louisiana at Monroe), 20; *Monroe Morning World*, January 1, 1946.

[54] Interview with Earnest Strauss, Monroe, LA, January 31, 2005. The theater's origins date back to a post-war play that Mr. Strauss attended in company with Mr. and Mrs. Clifford Strauss. It was a cold winter night and snow was falling. Indeed, a tear in the roof allowed the snow to fall on the stage behind the curtains. Nevertheless, the play went on. On the way home later that night, Clifford decided that this was ridiculous and offered to donate $60,000 toward the construction of a permanent theater with the proviso that it be matched by the community.

CHAPTER II

Navigational Training

The cadets found navigational instruction at Selman to be a very stressful experience. It involved equal parts classroom study and physical conditioning. Even though the basic course had been created by Colonel Harbold ten years earlier, the program was revised, strengthened, and lengthened as conditions warranted. For example, instrument navigation came of age during this time and was incorporated into the training. Weather forecasting and radio communications also received additional emphasis. To assure a steady supply of knowledgeable instructors, an instructor-training program was also established.

In order to select the best possible student navigators, the Army Air Corps had established a classification center in Nashville, Tennessee. It was there that potential students were thoroughly screened. The process determined who would be directed into pilot, bombardier, or the navigator training program at Selman. Rigorous calisthenics, courses in military protocol, and guard duty occupied the hours between testing. However, there was ample down-time during which the cadets played cards or read. The candidates underwent a complete physical examination and a complex battery of psychological and general academic tests. They were required to have a fundamental knowledge of reading, writing, and mathematics. Initially, a concerted effort was made to select for navigational training only those with some prior college experience, but due to the demands for trained navigators early in the war, this requirement was replaced by rigorous qualifying tests at the classification center.

The physical exam was thorough and decisive for being accepted into the navigation program. The men were tested on their motor coordination, response time to different stimuli, pulse rate and blood pressure before and after strenuous exercise, and overall physical condition and dexterity. There was an especially extensive array of eye examinations and in particular exams for hand-eye coordination. Tests were conducted for color-blindness, convergence, field of vision, as well as depth perception. Many would-be navigators were fitted with lenses to correct their visual impairments. Those who washed out were quickly funneled into other aerial training such as gunnery, radio operator training, or flight engineer training.

The psychological profile of the future navigators was also of paramount importance. A battery of psychological tests was intended to identify those individuals best suited for navigational training through their prior experience, educational background, and aptitude. The ideal candidate had to possess a high measure of mental and physical resiliency. However, the most significant psychological component was deemed to be self-confidence because self-confidence was conducive to victory. The navigator needed to be resourceful. Although he had to be sure of his abilities, at the same time he needed to recognize his shortcomings, secure in the knowledge that he could overcome any obstacle.

The would-be navigators were also tested for their ability to act under pressure. They had to make the right decisions under tremendous stress in a limited amount of time. No matter the distance to target, the weather, or the ferocity of enemy opposition, the navigator had to furnish information that would bring everyone safely back to base. No matter how frightened he might be in combat, he had to act calmly and capably. Even if the plane encountered extreme turbulence or had to escape enemy pursuit, or anti-aircraft fire, the navigator had to make accurate observations from which he could draw correct inferences. He had to be able to spot barely recognizable landmarks several thousand feet below on the ground, recheck his maps and charts, and do the mathematical calculations based on the half dozen tables necessary to make the correct deductions. There was no room for error; the success of the mission and the safety of the plane and its crew depended on the navigator.

The mission of Selman Field was to turn out trained navigators to serve as crewmen in bombers, cargo, and transport planes. To do so, the military created two schools: a pre-flight and an advanced navigation school. The normal school day for both was equally long and grueling: 5:30 A.M. until 11:00 P.M., Monday through Friday. The pay was $75 a month.

The pre-flight program lasted nine weeks and consisted of equal parts physical exercise, military training, and academic coursework. Although academics was a priority, physical conditioning was still important. Drills, parades, and physical activity were all included. In order to design the exercises, studies were conducted of the muscular motions made by a navigator moving through a bomber. These exercises were designed to teach the navigators the limits to their mobility under various combat conditions and which muscles were used under different circumstances. The day began with a one-hour run; the cadets would run double-time in platoon formation while they sang songs to mark the rhythm. Sometimes, following the run, they would do calisthenics, as the staff leader in charge of physical training shouted commands. At least once a week there were cross-country seven-mile runs, often with obstacles placed on the "jogging trail."[1] A schedule of intra-squad competitions also helped keep the men in top physical condition.

Although the training missions did not reach high altitudes, the students were taught to respond to the dangers of flying at high elevations. Since normal breathing was difficult, if not impossible, due to the low air pressure and lack of oxygen at such heights, the crew had to use oxygen masks. However, the navigator was the only crew member who had to move around during flight. Therefore, at each of his various posts he would find a mask; but as he moved from one post to another he would have to manage without a mask. This meant that he risked dizziness or fainting if he lingered too long between posts. Yet, no matter what, the navigator had to remain mentally alert and make accurate readings in order to ensure the safety of the crew and the success of the mission.[2]

Military training consisted of daily close-order drills. The cadets were organized into flight squadrons; they marched in formation to both class and the mess for meals, as well as in periodic parades. In basic military science instruction, the cadets studied the organization of the Army air forces, ground forces, and supply organizations. They learned to identify aircraft by nationality and type. Other areas of study included knowledge of court martial and military boards, command and staff functions, chemical warfare, uniforms and insignia, discipline, and military conduct. Cadets also learned how to use Morse Code.

Academic instruction stressed meteorology, physics, and mathematics. Classes began at 7:50 A.M. and continued until 11:45 A.M. when lunch was served in the mess. At 1:00 P.M. classes resumed and continued until 3:45 P.M. Evening classes were also held and the classrooms were always open for study, often with an instructor assigned to each classroom to tutor and work with the slower students. Lectures were supplemented by training movies and weekly exams.

The study of meteorology involved weather and climate; the fundamental elements of weather included the study of air pressure, temperature, and humidity, isobars, isotherms, weather-map symbols, and the identification of air masses and fronts. Physics instruction focused on the structure of the atmosphere, air as a mass capable of supporting weight, airfoil shapes, the angle of attack, lift forces, wing design and structure, and tail surfaces and general plane design. Math was especially important because of the demands of the highly technical courses. Indeed, pre-flight instruction included forty-two hours of intensive mathematics, almost the equivalent of a college semester. The cadets were also taught basic calculations for different types of navigation. They were made comfortable with the sextant, calculator, and other navigational tools used to determine position in the sky. They also were exposed to the pressure chamber to determine whether or not they would be able to fly at high altitudes. On the day of their testing they were placed on a special diet of non gas-producing foods. They were then put in the pressure chamber for three hours. Once in the chamber, as the pressure in the tank decreased, the gases in their bodies expanded and the cadets had to swallow to equalize the pressure. This was equivalent to flying in an unpressurized bomber at 38,000 feet.

The only departure from the strict schedule was on Saturday when the cadets had to clean the barracks and latrines thoroughly in order to pass the formal "white gloves" inspection. When the inspection was over, the squadrons competed in an intense marching competition for awards and recognition; as they marched, the band played John Phillip Souza's "The Stars and Stripes Forever."

After successfully completing the pre-flight program, many cadets were assigned to the six-week aerial gunnery school in Florida, followed by the fifteen-week advanced navigation program back at Selman. In April 1943, this was extended to eighteen weeks. Here only one subject was stressed—navigation: how to get the bomber to its target, knowing the plane's location at all times, and how to return it safely to base. Before the twentieth century, the science of navigation had been confined to travel at sea, but the advent of aerial warfare had given it new dimensions. Exact positions had to be fixed while the plane was flying at speeds of upwards of 300 miles an hour, all while fighting off enemy attacks and anti-aircraft fire. Moreover, the navigator had to maintain his bearings while crossing vast expanses of water or while clouds covered key landmarks. In addition, the navigator had to be certain he had enough fuel left for a safe return.

Navigational training included four different procedures to plot location. The first, pilotage, utilized landmarks and involved an intensive study of maps. This technique had serious shortcomings, especially over large expanses of water; normally islands could serve as points of identification, but overcast skies could interfere with the ability of the navigator to observe them. Cloud cover was equally a problem over land. Moreover, especially in the early stages of the war, maps and charts were not always accurate and did not correctly identify all important landmarks.

Dead reckoning was the second type of navigation. This method tracked direction and distance from the point of departure. It depended upon accurate instruments, which measured the plane's motion as well as the speed and wind drift. The cadets were also taught to calculate wind speed and direction by observing wave motion on the water. To make certain course and direction were accurate, the navigator had to maintain a meticulous log of the trip.

Radio was the third type of navigation. It consisted of "riding the beam" from one radio station to another until the plane reached its target. However, there were often no radio stations in proximity and the navigators were therefore without beams to guide them.

The final method was celestial navigation. This proved to be the most difficult method for the cadets to learn. It depended upon the accurate identification of stars and other celestial bodies. The students were required to find a plane's position by sighting or "shooting" at a star, the sun, or a planet using a sextant to find the altitude of that celestial body. Then they recorded the altitude of three of these bodies, while referring to a periodical called the *Air Almanac* as a reference. Using the information gathered, they had to develop a "line of position," or LOP. By shooting three LOPs from three different sources and drawing a triangle, the students were to locate their position relative to the ground at the point where the lines intersected. These calculations took roughly ten minutes; but in that time the plane had moved a considerable distance. Therefore, the students had to project their trajectory from the last known position through their fix. They had to factor in the plane's speed, wind drift, and heading, and consequently make the necessary adjustments in order to reach their destination.

In addition to these approved forms of navigation, many cadets also relied on other less conventional methods to find their way. Some actually assessed their proximity to Selman Field using their sense of sight, relying on visual contact with the huge plume of smoke from the West Monroe paper mill. There were incidents, however, where the cadets confused the smoke coming from the Bastrop paper mill, located roughly twenty-five miles to the north, with that of the West Monroe mill. Still others used the red beacon on top of the downtown Francis Hotel to guide them back to Selman.[3]

Classroom instruction for both schools took place in rooms that seated forty to forty-five students. In front of the room, on a raised podium, stood a lectern and a large blackboard. Around the room were a number of tables used for plotting. The advanced navigation school always had the latest available equipment for hands-on instruction. Each student was given an octant, a navigational instrument used to determine the exact altitude of a star at a precise time. The cadets could be awakened at any time between taps and reveille; they were then ordered to practice "shooting the stars" with their octants. The students were also given an E6B, a kind of circular slide rule to help calculate distance and determine wind drift, the amount by which the wind was blowing a plane off course. In addition, all students were given watches and chronometers, which gave the precise time. All students were required to record and maintain accurate logs of their missions. The goal of each student was to arrive at "zero, zero," which meant hitting the target at both the precise location and at an exact time. Even one inch off target was considered a failure and was unacceptable.

The instructors, military and civilian alike, were knowledgeable and businesslike. They employed both lectures and training films to present the information to the cadets. Exams covering the week's instruction were given at the end of each week. If needed, classrooms were opened for additional study time. A number of simulators were also used during classroom instruction to improve the students' navigational skills. A dead reckoning trainer, a wheeled vehicle with aircraft instruments, helped the students become familiar with procedures and gave them practice maintaining accurate logs.[4] The Link celestial navigation trainer was a huge cylinder designed to simulate conditions for the combined training of the pilot, bombardier, and navigator. At Selman, the cylinder was modified to train only the navigator. It helped the students become more comfortable with pilotage, dead reckoning, and celestial navigation. Dr. Edwin A. Link, inventor of the trainer, visited the post on several occasions to offer his personal recommendations.[5]

As the war progressed, additional changes were made in the curriculum to reflect combat experience, and the introduction of more recent discoveries. The highly successful British "G" system was incorporated into instruction in 1943. But it was soon outdated by rapid developments in technology and replaced by the American-invented "Loran" system in 1944. If reception was good, an electronic box accurately determined the plane's position to within thirty feet in microseconds by measuring the time differential for radio signals from two different stations in reaching the receiver. The fixed position of the plane was determined by where the signals intersected.[6] Then, in May 1944, a newly-developed sextant, the B-12, was introduced and issued to each navigator. It virtually eliminated error in determining latitude.

The final phase of student instruction, which took place both day and night, involved one hundred hours of aerial training. It was at this time that students received practical experience in the theory they had studied in the classroom and could finally apply what they had learned. Most of the training missions took place in a specially equipped AT-7 twin-engine Beechcraft which had a top speed of 225 miles an hour. It accommodated the pilot, a certified navigator, and three or four students who were seated in desks behind the cockpit. Each desk was equipped with a complete set of instruments, including a drift meter, an altimeter that gave altitude, and an airspeed indicator that registered the airspeed but not the ground speed. A second plane, the Lockheed AT-18, was also modified for student instruction. It had a longer range and faster speed than the AT-7. In addition, it had a clear dome in the top of the fuselage that allowed the cadets to "shoot" celestial bodies with their octant. Late in the war, the students were required to fly at least one high-altitude mission while wearing an oxygen mask to simulate wartime flight conditions.

During four- and eight-hour flights across a half dozen states, the instructors constantly bombarded the students with problems that they were required to solve. The students were expected to use their instrument training to take corrective measures. These measures were then analyzed by both the instructor and students and the cadets were graded. All training missions consisted of a triangular flight pattern of three legs, beginning and ending at Selman. The cadet in the first seat was designated the lead navigator for the first leg. He was to use celestial techniques. If it was a day flight, he would use the position of the sun. He told the pilot which course to fly. The second student navigator was responsible for "dead reckoning;" he used his compass and wind drift that was determined by a driftmeter, as well as airspeed and time to plot his course. He was warned not to look out the window during the flight. The third student navigator was to use "pilotage" techniques, using charts and maps and comparing them to ground features to plot the course. The lead navigator was required to advise the pilot as to the plane's ETA. The cadet was also expected to tell the pilot to make course corrections due to wind drifts. At the end of the first leg, the pilot would tell the lead navigator whether he was correct; if he was not correct, he would inform him how many miles he was off course or how many minutes he missed his ETA. The other two students would compare this information with their calculations to determine by how accurate they had been. On the second and third legs, the other two student navigators would respectively become lead navigator and follow a similar procedure.[7]

The last or graduation flight was eagerly awaited by all the students; it was a night mission testing the students' knowledge of celestial navigation. The cadets were rewarded for their efforts by being allowed to spend the night wherever they had flown. This permitted the students to have a night of liberty, free from military restrictions, before they had to return to Selman Field the following day. Those students who did not receive a passing grade were usually remanded to the next session to receive additional instruction. With their instruction in the "flying classrooms" complete, the students were measured for their new uniforms and received their navigator wings. They were commissioned as second lieutenants or flight officers.

Unfortunately, however, in spite of the effort to thoroughly screen all applicants, the demanding nature of the training and the stringent requirements at Selman Field took their toll. Cadets could be washed out if they became airsick more than six times. If they could not absorb all the instructional information, they could be dropped back a class or two. Only about half of the cadets managed to satisfactorily complete the course and become commissioned. The rest were reassigned to military duty elsewhere.

Endnotes

[1] *Selman Field: World War II: 1942-1946* (Paducah, KY, 1998), 28.

[2] Robert Hallack, *Selman Field: An Army Air Forces Navigation School* (1984), Special Collections, University of Louisiana at Monroe, 7.

[3] Norris B. Harbold, *The Log of Air Navigation* (San Antonio, TX, 1970), 97.

[4] Ibid., 97-98.

[5] *Wings Over America: United States Army Air Forces* (Baton Rouge, LA, [1943]).

[6] Harbold, *The Log of Air Navigation,* 98.

[7] *Selman Field,* 28.

CHAPTER III

The Missions

Following graduation from Selman, the newly-minted second lieutenants reported to combat duty around the world. I have collected a number of their stories, based on oral interviews, e-mails, telephone calls, and letters from the former airmen themselves or their families. Their recollections are mere hints of the real measure of their courage, anxieties, hopes, and fears in a war with little mercy. These men represented a broad cross-section of American life. They came from large cities as well as sparsely-inhabited rural areas throughout the nation. Many had not traveled beyond their home counties, let alone outside of their home states. All experienced the dire circumstances of the depression years. Most of them had never been in an airplane. Yet, the common denominator that united them was their desire to fly. They were trained at Selman Field to perform a task requiring a high degree of mental and physical coordination under extremely difficult and dangerous circumstances. They left behind the certainties of their previous lives to meet the challenges that confronted them head-on. To survive the horrors of war, they had to develop a thick outer hide, a tough attitude, and increased resourcefulness.

Though it is unlikely the individual actions of most of these men would be cited in history books, they, nevertheless, deserve to have their deeds recorded. With each passing year, World War II becomes even more remote in public memory. Yet, future generations should be aware of the role the Selman Field navigators played in determining the war's outcome. By relying on their training to overcome obstacles, these men became extraordinary in the true sense of the word; as Tom Brokaw said, they really were "the greatest generation." These few accounts are offered as a tribute to all those heroic navigators who fought and died in order to preserve our freedom.

Harry Peterson

Harry Peterson, from Rock Island, Illinois, passed through the Nashville classification center and was assigned to Selman Field for navigation training. Thrilled that he would be able to fly, he boarded the train for Monroe. However, due to the train's late arrival, Peterson missed his connection in Meridian, Mississippi. After waiting hours in the stifling heat and humidity, he boarded the next train. The passenger cars were hot and cramped. The windows were kept open for ventilation, but the soot from the coal-fired engines covered the passengers, making it difficult to breathe. Peterson arrived in Monroe late at night; his uniform was soaked with sweat, and he was exhausted. He wondered if he had made the right decision.

Peterson began training at Selman Field as a twenty-year-old in 1943. He graduated as a second lieutenant with his navigator wings in February 1944. He was then ordered to report to Fresno, California, where he joined up with his crew. From there, they flew in a B-24 to Hawaii for additional instruction. Training missions allowed members of the crew to practice their individual assignments; at the same time they allowed the airmen to coalesce into a cohesive fighting unit and gain confidence in one another. This bonding was critical for their survival, as each had to depend on the others to perform their tasks correctly. They also had time to become accustomed to the B-24, an extremely cramped aircraft. Peterson had to bend down and carefully edge into the bomber through the nose wheel-well of the plane. He then wedged himself into a tiny compartment behind the flight deck; there he sat on a metal stool in front of a shelf holding his charts.

When the crew completed their training, the members were assigned to the 7th Air Force, 30th Bomb Group, at Saipan in the Marianna Islands. Their B-24, named "This Love of Mine," had a picture of a glamorous, scantily-clad girl under the pilot's window. She served as the mascot for the entire crew.

Saipan had been seized during fierce fighting in the summer of 1944; signs of the bitter struggle were still evident when Peterson arrived. The new arrivals were expected to complete thirty-five missions before rotating home. Their first missions were to Iwo Jima, which was being riddled by daily bombing runs prior to the impending invasion. Concentrated aerial assaults against the two airfields and defensive targets were non-stop. Peterson and his crew were ordered to neutralize the Japanese fighter planes on the base by dropping delayed-action fusion bombs to maximize damage and keep the Japanese off balance.

The first few missions were uneventful. While Peterson and his crew did encounter antiaircraft fire, it was not accurate and they returned safely to base. Conditions on the plane during the bombing runs were cramped, cold, and uncomfortable; in addition, the food the crew carried aboard was unsatisfactory, usually unappetizing "C" or "K" rations. A few times, the men were fortunate enough to receive canned turkey; Peterson carefully spooned it out onto bread for a sandwich. This proved to be one of his favorite war-time meals.

The successful completion of the early missions did not lessen the stress or sense of danger for those aboard the plane. Each time they took off, they realized they were in a metal cylinder, little more than an airborne bomb filled with gasoline and explosives; they could all be killed in an instant, whether because of enemy fire, human error, or mechanical failure.

Peterson was not exempt from the extreme anxiety they all felt each time they lifted off. Nevertheless, he tried to prepare as thoroughly as he could for each mission. From the flight plans he was given at pre-flight briefings, he drew all his courses on the sectionals and other maps. Then he compiled all the navigational information, including weather reports and expected enemy opposition. His summary covered all aspects of the operation. He then went to the plane, checked oxygen and guns, and briefed the crew. When he had finished, everything was as ready for the mission as possible. With eight missions behind them, Peterson and his crew found the bombing runs had become somewhat routine. Their next mission, however, proved to be anything but. Peterson was seated behind the pilot as the bomber approached the target. He had just gotten up from his seat when an anti-aircraft shell exploded right beside the plane; he felt a severe jolt as shrapnel tore through the plane, shredding the seat he had just left. Afterwards he never sat in the navigator's chair again, preferring to sit on his helmet at a lower level next to the navigator's desk by the open bomb bay. Whether it was luck or fate, Peterson stuck to his new routine and survived the war unscathed.

Upon returning to base, Peterson had ample time to reflect on his good fortune; there was little else to do on Saipan but think. Sometimes the crew would spend the day after a mission sleeping anywhere from twelve to fifteen hours. On succeeding days, they would usually play cards: poker or cribbage. They would joke, laugh, and talk about their missions, all the while feeding off the group dynamic and building up their morale. At other times, they would walk on the beach or explore the island. The island was small, twelve miles long by five miles wide; brisk walks took up a good part of the day and helped pass the time. Toward the end of the tour, during the last few missions, partying became the main safety valve for stress release. The aircrews would enroll in a "whiskey lottery:" as a plane took off, they would try to guess where it was headed. The lucky winners shared a bottle of whiskey as a reward.

On one raid, Peterson's B-24 carried a load of 100-pound cluster bombs. The ground crew on Saipan had fastened the bombs together with snap clips and placed them on racks in groups of three. When the bombing run began, two bomb clusters on the lowest rack got hung up on the bomb-bay doors and the bomb above tumbled over the others on the rack. The three bombs caught on the cable that closed the bomb bay. Their propellers spun off and the armed bombs swung precariously back and forth. Peterson was the first to notice. Had he not taken his new observation post next to the open bomb bay, he would not have been aware of the impending calamity; if the bombardier tried to close the bomb-bay doors, the plane would have exploded and the crew would have simply been listed as missing in action.

Peterson knew that the slightest jarring would cause the bombs to explode. He immediately notified the other crew members. All recognized the urgency of the situation; they realized the bombs had to be jettisoned. The bombardier, a full-blooded Dakota Indian called "Tank" by the crew, rushed to the open bomb bay to assist Peterson. Neither man had time to put on his oxygen mask or parachute. Despite the rush of cold air, the two men climbed out onto the crosswalk, cautiously squeezing between the bomb racks. They carefully made their way along the narrow catwalk, and while Peterson clung to Tank's leg to steady him, "Tank" gingerly unhooked the bombs from the cable with his left hand. Then he did the same thing with the other bomb cluster.

On another mission over the island, Peterson witnessed a Japanese suicide plane fly directly into a B-24. The result was a huge explosion. Only the tail gunner survived, though he was badly burned. The sight deeply disturbed Peterson and his crew. This was another reminder of how important it was to remain vigilant.

After landing, the shaken crew member were taken to their routine debriefing where Peterson and his men were given rye whiskey; this helped them calm their nerves and relax so they could better relate the details of the mission. However, the traumatic memories became recurring nightmares for many years after his discharge from the service.

There were many other close calls. While Peterson was on Saipan between missions, other crews continued the daily assaults. One plane, captained by a friend of Peterson's, "Ma" Faigan as he was known around the island, was completing a mission over Iwo Jima with a replacement crew when his plane was shot up by intense and accurate FLAK.[1] Faigan put the plane on automatic pilot even though the plane had lost its rudder, had a crippled hydraulic system, and had a nose wheel that could not be lowered. The crew voted to try to make it to Saipan rather than bail out. In spite of the crew's efforts, the plane went down in the ocean between Tinian and Saipan. When Peterson and the others heard the news, they feared the worst. However, while they were having dinner, they learned that "Ma" Faigan and most of his crew had survived. Their Mae Wests had inflated and, when the plane broke up in the water, the men floated out. Only the tail gunner and radio operator had been lost.

Perhaps the most significant achievement of Peterson's missions was the accidental discovery of the jet stream. The round-trip flight from Saipan to Iwo Jima was long and tedious, usually thirteen to fifteen hours. As always, Peterson calculated air speed and time to target. However, on one attack, his bomber formation altered tactics and approached Iwo Jima at an altitude of 28,000 feet from the western side of the island. Peterson looked at his maps, did his calculations, and surmised that air speed increased significantly because of the air stream at that altitude and approach. At the debriefing following the flight, he reported this to his superiors; thereafter bombing runs followed this new course with increased speed and accuracy. Prior to that time, the air speed had been much slower and many of the bombs had fallen harmlessly over water.

Peterson eventually flew forty missions between August 20, 1944, and December 31, 1944. Halfway through the missions, he began to consider what he wanted to do in the future. He decided to go to college on the G.I. Bill. He went on to a long and rewarding career, first as a junior high school history teacher for sixteen years and thereafter as a junior high school principal for twenty-two years. He raised a large family of seven children, and today has nineteen grandchildren and three great grandchildren.

As Peterson looked back on his time at Selman Field, he did so with nostalgia and a deep sense of gratitude. He reported, "Selman Field and navigator training prepared me for life and made me into a man."

Jeff Hawkins

Jeff Hawkins was born and raised in Birmingham, Alabama. After induction into the Army and basic training, he was sent to the Nashville classification center. Then, in the summer of 1943, he was ordered to Selman Field. After finishing pre-flight school, he went to gunnery school at Tyndall Field in Panama City, Florida, for a six-week course. There the men learned to tear apart .50 caliber machine guns, properly clean and lubricate them, and then put them back together. The training culminated in a darkened room where the airmen were given assembled guns and told to strip them down. Then the instructor passed through the room in the dark to feel the parts and ensure that the weapon was completely disassembled. Afterwards the men had to reassemble the weapon, cock it, and snap the firing pin to show that the procedure had been done correctly. Later Hawkins learned to fire small arms, automatic weapons, and heavy machine guns from moving platforms. He then learned how to sight, swing, and fire twin .50 caliber guns from airplane turrets. The final phase of training involved aerial firing of machine guns from B-24 Liberator combat bombers.

Since the B-24 flew at high altitudes, the men had to climb from turret to turret and fire while encumbered by oxygen equipment. The targets were long canvas sleeves towed on cables behind AT-6 training planes. Each student's ammunition was painted a specific color and the students were graded by the number of holes shot by bullets in their color that had hit the sleeves. Hawkins completed gunnery school and was then qualified to wear the silver wings of an aerial gunner. Afterwards, Hawkins finished advanced navigation training at Selman Field in July 1944. Shortly before training was completed, he was married in the base chapel in June 1944.

Following advanced navigation school, Hawkins was assigned to Ardmore, Oklahoma, where he met with the rest of the replacement crew for the B-17 on which he was to serve. The new crew was then dispatched to Omaha, Nebraska, where they picked up their B-17. After flying to Great Britain in October 1944, they were assigned to the 34th Bomb Group, 391st Squadron, 8th Air Force.

Weather permitting, Hawkins and his crew began their almost daily aerial assaults. During their missions over Germany, the targets varied: oil storage facilities, marshaling yards, and industrial sites. The pressure on Hawkins remained constant. His task as a navigator was to know exactly where his ship was at all times. He had to be absolutely certain to reach the target, but his mission was always made more difficult by heavy FLAK from enemy fighters as well as poor weather conditions. Therefore, his preparations for each mission had to be thorough. He listened attentively at pre-flight briefing sessions, knowing that the safe return of his plane depended upon his actions.

On Hawkin's third mission, his plane was in the second group of the bomber stream. As it approached its target over Berlin, it encountered intense and accurate anti-aircraft fire. The plane shuddered, as FLAK ripped through it, knocking out power in two engines. The plane's captain feathered the propeller. As the plane lost power, it fell out of formation. However, the pilot was somehow able to keep the plane on course and the bombs were dropped over the targeted oil storage facilities. On its return flight, losing altitude and speed, the plane dropped further and further behind the formation. Over the North Sea, the plane was intercepted by ME-109s that fired rockets at it. Hawkins grabbed a machine gun and began firing back. For whatever reason, the enemy planes veered off and allowed the crippled bomber to return to base. Fortunately, Hawkins and his crew had no other encounters quite that close during the remainder of their tour of duty and Hawkins was able to complete his thirty-five missions. When the war ended, Hawkins returned to New York City aboard the *Queen Elizabeth.*

Sam Humphrey

Sam Humphrey, from Canton Center, Connecticut, joined the Army in 1942; he was trained to be a weather observer. In 1943, he was sent to Buckly Field, Colorado, for classification. After passing the physical and mental tests, he was dispatched to Montana State College, where he spent the next five months. In April 1943, Humphrey went to Santa Anna, California, for pre-flight school. Although the curriculum was demanding and a large percentage of the class washed out, Humphrey finished pre-flight successfully; however, all slots in the navigation schools were full. This meant that Humphrey and the rest of his class had to wait two months for openings. Humphrey worried that "spending so much time in training would keep me from ever making it to the war."

Then, when the wait was up, in October 1943, Humphrey was finally directed to Selman Field. When he arrived, he was informed that the navigation program had been reduced from twenty to sixteen weeks due to the high demand for navigators. However, because of the harsh winter weather, many of his training missions had to be scrubbed. As a result, the course was extended one week. Seventeen weeks later, Humphrey graduated; but even though every preceding class had received orders upon graduation, Humphrey's class received none. Humphrey continued to fear that the war would be over before he could be a part of it.

During a fifteen-day leave, Humphrey went home to marry his childhood sweetheart, Mary Mills. Their return trip to Monroe was to serve as their honeymoon. After their arrival, they rented a room in a hotel and immediately began to look for a place to rent for two to four weeks. Humphrey saw a group of men on the steps of the Methodist Church in downtown Monroe and asked them about short-term lodging. One of the men suggested he go to Riverside Drive and see a Mrs. McReynolds. He said that she was interested in renting a room to one of the local servicemen. So the Humphreys boarded a city bus and went to the McReynolds' residence. She gladly rented them a room in her home, welcoming them warmly. What the gentleman on the church steps had neglected to tell them was that he was Mr. McReynolds. Humphrey later recalled: "we were still there when he came home. And we were going to get the bus back to the hotel and get our bags and come back up, and he reached in his pocket and took out his car keys and handed them to me. By then it was dark and he said: 'Here, take my car.' Well, I was thunder struck.... I said, 'but you don't know me.' 'You are a nice family', he said. 'Well, that's good enough. You do it.'"

The Humphreys settled in, enjoying the gracious hospitality extended by the McReynolds. However, the war was never far from their minds. They learned that the McReynolds had lost a son, Sam Jr., a West Point graduate, in the desperate fighting for Corregidor in 1942. After only a month, Humphrey received orders that he was being shipped out to the southwest Pacific. Together with 200 other members of his Selman Field graduating class, he was told to report to the overseas replacement depot in Kearns, Utah. After a week at Kearns, he was sent to Travis Field, California. There he was issued a bayonet, and together with eight other airmen, crammed into the bomb bay of a B-24. They spent thirty-nine hours in the air en route to Nadzab, New Guinea. The uncomfortably long flight was relieved only by brief refueling stops in Hawaii, Tarawa, Manus, and Biak in the Dutch East Indies. On his first night in Biak, while he was trying to recover from the exhausting trip, Humphrey was greeted by a Japanese air raid. His wish for combat had become a reality.

Upon arriving in New Guinea, Humphrey was sent for seven days of jungle warfare training. After surviving this strenuous week, he was issued a .45 caliber pistol and fifty rounds of ammunition. He was assigned to the 70th Troop Carrier Squadron. For the remainder of the Pacific war, Humphrey moved north with the allied advance. He spent much of his time at an improvised airbase on Leyte and at Clark Field in Luzon in the Philippine Islands. From there he served as navigator on C-46s and C-47s, flying numerous support missions and dropping supplies for guerilla forces behind Japanese lines. On one occasion, he flew seventy Japanese prisoners back to Leyte from Minaleou.

On his most perilous mission, Humphrey volunteered to be the lead navigator for a glider assault on two Japanese-held airfields. The squadron commander, Jim Mize of Ruston, Louisiana, routinely refused to take a navigator on raids behind enemy lines to avoid exposing him to danger. Navigators were considered too critical to the success of the air war to be exposed to enemy fire unnecessarily. However, this raid called for pinpoint accuracy. Each plane was to tow two gliders, each loaded down with essential equipment to support the airborne assault force. Humphrey's two gliders were packed with Jeeps, howitzers, and munitions. He was ordered to make the drop at 600 feet. Coming in low and without warning, the Americans caught the Japanese defenders by surprise, and after a brief but bitter firefight, the airfield were secured.

In August 1945, Humphrey's squadron transferred to Ie Shima, a small outcropping twenty miles off Okinawa, where Ernie Pyle, the famed war correspondent, had been killed by a sniper. Shortly after the squadron's arrival, a massive Pacific typhoon smashed into the island, causing major damage to supplies, planes, and ships. After repairs were made, Humphrey and his crewmates flew troops and equipment from the Philippines to Okinawa in preparation for the impending assault on Japan. The ten-hour round-trip flight was made daily for a week, an exhausting pace for even the best aircrew.

After the atomic bomb was dropped on Hiroshima, the squadron was dispatched to Iwo Jima and ordered to carry gasoline to Yokohama to refuel the transports ferrying the occupation troops. At the time, it seemed possible that the fanatical enemy might attack again. Despite this concern, Humphrey was virtually unarmed except for one submachine gun and the .45 pistol each crewmember carried.

Humphrey made two flights to Japan before the surrender became official—for which he and his crewmates were awarded the Air Offensive Japan Campaign Star. After the surrender, Humphrey continued to ferry supplies to the occupation forces in Japan. Perhaps the assignment closest to Humphrey's heart during this concluding phase of the conflict was transporting gaunt and sick Allied war prisoners out of Japan.

One last assignment for Humphrey occurred when two C-46s from Humphrey's squadron were ordered to Hiroshima to measure radiation levels. With Humphrey serving as navigator, these planes were the first ones to land at Hiroshima following the dropping of the atomic bomb. The planes carried so much equipment that it took two days just to unload them before testing could begin. For his efforts, Humphrey was later awarded five campaign battle stars as well as the Air Medal.

During the Korean War Humphrey was recalled to active duty. He ended up spending his entire career in the Air Force, retiring as a lieutenant colonel. He and his wife Mary had been married for sixty-one years before she passed away in March 2006. They had three children and seven grandchildren, who together with Humphrey, put in a total of seventy-two years of active duty in the military service.

Bruce Knight Bockstanz

Bockstanz, from Grosse Pointe Park, Michigan, entered Oberlin College in the fall of 1940. He completed two years of coursework before being inducted into the Army Air Corps in July 1942. He managed to complete the first semester of his junior year before he was called to active duty in February 1943. He was assigned to basic training at Keesler Field, Mississippi. From there he was sent to the classification center at Nashville, where he successfully passed the battery of strenuous tests. He arrived at Selman Field for pre-flight training in July 1943, to be greeted by oppressive humidity and heat topping 100 degrees.

After successful completion of pre-flight training, Bockstanz, like Jeff Hawkins, entered the six-week gunnery school at Tyndall Field, Panama City, Florida. There he practiced skeet shooting with a shotgun and did target shooting with a .45 caliber pistol. He learned how to field strip and disassemble the many intricate parts of the Browning .50 caliber machine gun. He also practiced flying as well as shooting at targets that were towed behind a B-24. From gunnery school, Bockstanz returned to Selman for advanced navigation training where he immersed himself in his studies. When he graduated, he was ordered to report to the 393rd Combat Crew Training School in Sioux City, Iowa. There he met and immediately began training with his new crew aboard a B-17. After a month of flight operations, Bockstanz and his crew were sent to Kearney, Nebraska, to pick up their B-17 and fly to England. Shortly after takeoff, a fire broke out in the Number 3 engine, forcing the plane to make an emergency landing in Omaha. Following repairs, the crew took the same plane to England. They arrived on June 2, 1944, Bockstanz's twenty-second birthday. Bockstanz and his crew were assigned to the 96th Bomber Group, 339th Squadron, 8th Air Force.

The missions over Germany began almost immediately. On July 19, 1944, Bockstanz's plane was hit by a burst of antiaircraft fire and the crew was forced to bail out. Bockstanz avoided capture the first day but was taken prisoner on July 20. He was thrown into a lice-infested jail cell in Koblenz. After undergoing intense interrogation, he was transferred to a camp at Oberusel, northeast of Frankfurt. On August 5, 1944, he was shipped to Stalag Luft I in Barth, Germany. He remained there until Soviet forces liberated him in April 1945. Conditions were harsh, but at least when the first cool weather arrived, Bockstanz was moved from a tent to permanent barracks. There was even a stove that served the dual purpose of cooking and heating. In addition, the timely delivery of Red Cross food parcels helped him endure his months in captivity.

Ethel Kelly

Kelly, who was from Delhi, Louisiana, roughly forty miles east of Monroe, never served on active duty at Selman Field. Instead, she was one of more than six million women who worked in industry and shipyards during World War II.

Before going into the service, Kelly's husband had been a railroad man. Kelly felt that she too should contribute to the war effort. Soon after her husband's departure, she saw an ad in the Monroe paper that said Lockheed was hiring at its Burbank, California, aircraft factory. With little hesitation, the twenty-year-old Kelly, who had never been away from home, boarded a bus for the long ride to southern California. She went straight to the Lockheed complex and applied for a job. Although she had no prior experience working on an assembly line, Lockheed hired her and trained her to identify and rivet different metals. She then went to work as a riveter on wing sections of the P-38 fighter plane and later as an inspector on the assembly line. "I never missed a day during the war. I earned $1.10 an hour, worked the 4:00 P.M. to 12:00 A.M. swing shift, walked to and from work, and slept on a small bed in my room that I rented for $4.50 a week."

After the war, Kelly and her husband returned to Delhi. In 1991, she read in the paper that Selman Field veterans would be holding a reunion. Even though she had never served at Selman, she decided to attend because of her attachment to the military. At the reunion she met Rudolph Perez, who had been a mission pilot at Selman before he was sent overseas. Perez had fond memories of his time in Monroe and, indeed, his first son had been born while he was stationed there. After his discharge from the military, Perez had gone to work for Lockheed. He had worked his way up to vice-president, after thirty years with the company, before finally retiring.

At the reunion, during a casual conversation, they realized they had worked at the same Lockheed plant, even though not at the same time. After Kelly showed Perez her pay stubs and documents from Lockheed, he said to her: "Here we've been looking for a picture of Rosie the Riveter and a real Rosie the Riveter walks through our door." Perez immediately asked Kelly to be Rosie the Riveter for the reunion. From that point on, Kelly became the symbolic "Rosie the Riveter" for the Selman veterans.

General Rene Chesnais

Chesnais served as a lieutenant in the French Army during the German invasion of France. When German troops wheeled around the Maginot Line and plunged into France, Chesnais made a harrowing escape from German forces across the Pyrénées. He then managed to book passage on a ship leaving for the United States. He arrived at Selman Field in September 1944. He was one of more than 2,000 French cadets who went through training in the United States in a variety of schools: pilot, navigation, bombardier, and gunnery. A dozen of these cadets went through training at Selman; they included Chesnais, who left in July 1945, with more than 550 flying hours.

All of the French cadets, including Chesnais, instead of being sent to operational training units like their American counterparts, remained at Selman a minimum of four months after receiving their wings. They flew an average of one hundred hours a month in preparation for becoming future navigational leaders in the French Air Force. Routine training flights included long distance day flights and return night flights to test the cadets' skills at using celestial navigation. Some of the most difficult flights involved night take-offs from Selman Field. The navigator would remain blindfolded for two hours with earphones turned off so that he could not hear the pilot's conversations. Then, when the navigator was permitted to remove his blindfold, he had to chart the trip back to Selman using only a sextant, a watch, and his celestial charts. No electronic aids were permitted. The estimated ETA could be off by only one minute at the most. However, the navigators were also given training on the most modern navigational aids then available, including the Loran. This intensive training gave the cadets added confidence in their abilities.

Chesnais went on to a distinguished career in the French Air Force, retiring as a general. He remembered Selman Field fondly: "I have always said that Selman Field was the best Air Force School I have known, and I have known many around the world, and I am very proud to have gotten my navigator wings there."

Frank Serio

Frank Serio, from Monroe, Louisiana, served with the 8th Air Force in England during the height of the air war over Europe. From March to October 1944, he flew twenty nine missions, serving as lead navigator on twenty-five of them.

One of his most memorable missions occurred soon after D-Day in support of the French Maquis in the south of France. The Maquis were members of the French Resistance who waged a guerilla campaign against the Germans. They were always in need of additional supplies to continue the insurgency. Serio's plane flew at an elevation of 17,000 feet and was accompanied by P-51 fighter planes, which were flying support in case of enemy fire. As the plane approached the target drop, it gradually descended to 500 feet. The bomb bays were packed with a full load of canisters containing an assortment of munitions, explosives, grenades, and weapons. The crew was told to look for an open field, roughly one-fourth of a mile wide by one-fourth of a mile long. The field was to be demarcated by three huge fires arranged in a triangle. After a visual sighting of the target area from the bubble of their B-17, the crew flew over the field to scan for possible enemy activity; seeing none, the plane made a 360-degree turn with bomb-bay doors open. At the command "bombs away," the canisters were released.

As the canisters, strapped to parachutes, descended, the Frenchmen ran from the woods to recover the weapons, waving to the crew as they did so; the crew waved back. By the time the plane completed one more 360-degree turn above the field, Serio noticed that the field had already been completely cleared. Even the parachutes, together with the canisters, had disappeared. The crew flew home with a sense of satisfaction. Unlike other missions, this one was personal. Serio and his crew were able to see that their mission had been accomplished and had given added firepower to the Resistance.

Robert Brnizza

Robert Brnizza graduated as a flight officer from Selman Field in March 1944. He was then ordered to Casper, Wyoming, to meet his crew. Casper served as a retention center for crews before they were assigned to combat. In April, Brnizza was dispatched to Topeka, Kansas, to pick up a new B-24. From there, he and his crew began a long odyssey to Great Britain. They first flew to West Palm Beach, Florida, and from there to Trinidad, Belem, and Fortalaza on the northeastern coast of Brazil. Then they crossed the South Atlantic to Dakar, Senegal, and after refueling, flew north to Marrakech, Morocco. From Marrakech, they began the final leg of their journey to Wales. There they were to serve as part of the continual stream of replacements to the 448th Heavy Bomber Group, which had sustained major losses in the air war over Europe.

Brnizza's first mission came on June 6, 1944, "D-Day." He and his crewmates, though untested in combat, felt prepared. The pre-flight briefing emphasized the skill and tenacity of German fighter pilots, as well as the uncertain weather patterns: ground fog and other weather conditions that could obscure their vision and force them to rely on instruments. All of the men knew the risks they faced, but they believed their training and determination were equal to the task. In spite of heavy FLAK and fighter attacks over the targets, Brnizza and his crew prevailed. Other successful bombing raids followed. On June 21, they were ordered out on their fifth, and what proved to be their final mission. At 18,000 feet, directly over Berlin, the plane was hit by intense and accurate FLAK. When the order was given to bail out, Brnizza was the first one to jump. He delayed pulling the ripcord until he estimated he was at 5,000 feet. At that point the plane was hit again and went into a spin. The rest of the crew could not parachute out until the pilot finally regained control of the plane and managed to level it out.

Brnizza landed in a large park at the southern edge of the city. Wondering what had happened to the other members of his crew, he rolled up his chute and hid beneath the undergrowth in the park. Hoping that no one had seen him during his descent, he peered out from his hiding place; he saw motorcycle police riding up and down the street, obviously looking for him and the other airmen. Fortunately, he was not noticed. After night fell, he walked south in the park, trying to get away from heavily-guarded Berlin as quickly as possible. He was slowed down in his movements by his heavy flight boots, which sank into the spongy soil. The cumbersome flight suit he wore over his uniform to protect him from the extreme cold in the unpressurized bomber also hindered his movements. With no food, he knew he could not keep going for long; he would need nourishment and a safe house. He stayed hidden behind the tree line until he reached the edge of the huge park. After that he continued to walk south, staying hidden as much as possible. He tried to remain on the move so no one could get close enough to question him. However, after six days as a fugitive, he was finally captured. He spent the next ten months as a POW. After the war he was able to make contact with his pilot; he learned that six of his fellow crewmen had been shot to death while attempting to escape. No trace was ever found of the seventh crewmember.

Earl And Courtney Shankin

Earl and Courtney Shankin were fraternal twins; growing up in Chicago, they were inseparable. After graduation from high school, they entered the University of Chicago. Both excelled in athletics and were champion gymnasts. Earl won three NCAA gold medals in vaulting and Courtney won rope climbing and six all-around golds; they led the University of Chicago to the National Collegiate Gymnastic Championships in 1941. Their gymnastic feats even captured national attention. Both appeared in the July 1940, issue of *Life* magazine and Courtney's horizontal bar routine made the "PathMovie" News in 1941. Moreover, after their gymnastic season ended, both played on the university's varsity baseball team.

The twins enlisted in the army in January 1943. Following basic training, they arrived at Selman Field in April of that same year. In August, they completed the program and received their commissions. Following further training at Alamogordo, New Mexico, they received orders to report as navigators to the 450th Bomber Group, 722nd Bomb Squadron of the 15th Air Force in Italy. The operations area of the 15th Air Force was immense, including most of southern and eastern Europe: Italy, southern France, southern Germany, Austria, Bulgaria, Czechoslovakia, Hungary, Rumania, Albania, and Greece.

The Shankin brothers were an exception to the military rule that prohibited brothers from serving in the same unit. Ever since the deaths of the five Sullivan brothers, who had all been crewmates on one vessel when their ship was torpedoed in 1942, brothers were not assigned to the same unit. Nevertheless, the Shankins requested assignment together. Thanks to assistance from their uncle, a retired World War I Army Air Corps colonel, their petitions were granted.

The Shankins first served in Italy as navigators on separate planes; they only flew one mission together. In northern Italy, they were ordered to hit the railroad marshalling yards in Alessandria. Earl, who was the lead navigator, asked Courtney to fly as nose gunner and handle the target-sighting instruments. During the mission, an unrecognizable voice came over the intercom asking Courtney "how many minutes to enemy territory?" "Ten minutes, sir," he answered. The voice replied, "You don't have to call me sir, I am your brother." The words helped Courtney to relax. Earl later said that, "on that mission nothing happened. We met no fighters, ran into no FLAK, and hit the target right on the head." It was what the airmen thankfully called a "milk run."

Not all missions, however, went so smoothly. During one mission, Courtney's plane was forced by an oil leak to leave the formation over the Brenner Pass between Switzerland and Italy and turn back. Earl did not notice that his brother's plane was not with the others. Earl's plane returned to base in Italy six hours late as it was low on gas and had to land in Corsica to refuel. Each man thought he had lost the other until they were reunited that night.

On base, between missions, the brothers found little to occupy their time and keep their minds off their next assignments. Emotionally drained by six-to-eight hour missions and physically exhausted by constant stress, the brothers tried to rest and forget. Entertainment was limited, usually consisting of two movies shown outdoors twice a week. Good food was also scarce. The "C" rations they were served on base were bland and unappetizing. They were allotted only two candy bars a week, which did little to satisfy their cravings. On many occasions, therefore, they would travel sixty–five miles one-way to Bari, Italy, 15th Air Force Headquarters, to purchase ice cream and other sweets.

The brothers had another scare when they flew a bombing run targeting the Ploesti oil fields located thirty-five miles north of Bucharest, the capital of Rumania. Earl was in the "left box" of the eighteen-plane formation and Courtney was in the "right box." As they entered Rumanian air space, forty German fighter planes swarmed around the formation and succeeded in shooting down two American planes from each box. In the melee, each of the brothers feared the other was in one of the planes that had been hit. Soon, however, they sighted one another and waved, relieved to know the other was still alive. Earl was later awarded the Distinguished Flying Cross for his presence of mind and quick reactions during the mission.

Fifty was the "magic number" after which airmen were grounded and shipped home for reassignment. On the twins' forty-fifth mission on May 12, 1944, only five missions from the magic fifty, the bomber group took off for its target. Courtney was aboard his Liberator bomber, painted with a figure the crew called the "Booby Trap." The crew assumed Earl was aboard his usual plane in the formation. During the bombing run, the plane Earl usually flew in was shot down. The men watched in horror as the damaged plane disappeared from view. Over the intercom Courtney hurriedly reassured them that Earl was back at the base waiting to fly later that afternoon. The Shankins later learned that no one had survived that crash. Only seven days later, on May 19, Courtney's crew reached fifty missions and thus became the first crew to complete its tour of duty and the only original ten-man crew from the Alamogordo training base to finish without any losses.

The Shankin brothers earned numerous decorations because of their harrowing combat experiences. Courtney was awarded six Oak Leaf Clusters as well as his Air Medal, and Earl, five. Besides his DFC, Earl also received three Battle Stars while Courtney was awarded two. Both were given European Theater of Operations ribbons. After they were shipped home, the brothers were assigned to Selman Field as instructors. The brothers remained at Selman for the duration of the war, thus concluding their military careers where they had begun their training.

Following the war, the brothers remained in close contact with the other members of their crews. Courtney settled in Highland Park, Illinois, where he served for many years on the board of directors of USA Gymnastics, the governing body for the sport. In 1967, he coached the American team in meets against Canada and Mexico. Earl settled in Sausalito, California. Both men enjoyed successful business careers.

Peter William Sommer

Sommer enrolled in classes at Yale University in the summer of 1942. However, he had only completed his first semester when he was inducted into the Army in February 1943. He passed through classification and training at a number of bases: Nashville, Tennessee; Atlantic City, New Jersey; Springfield, Massachusetts; Montgomery, Alabama; Jackson, Tennessee; and Fort Myers, Florida. Finally he was dispatched to Selman Field for navigational training; he graduated with his wings in April 1944. From Selman, he passed through a staging area at Herington, Kansas. He then shipped out to Great Britain in July 1944, where he was assigned to the 7th Squadron, 34th Bombardment Group of the 8th Air Force stationed in Suffolk, England.

Sommer flew twenty-one event-filled missions in the air offensive against Germany. On January 20, 1945, Sommer's plane, a B-17, lifted off heading toward the assembly point for the formation over the English Channel. There he was to join the bomber stream en route to the target over Germany. Over the channel, the plane developed engine trouble and the engine burst into flames. The plane veered out of control and plunged toward the sea. The pilot, fearing an explosion, ordered several of the crew, including Sommer, to bail out. The airmen parachuted out south of Ipswich. In the meantime, the speed of the dive partially extinguished the fire; the pilot succeeded in bringing the damaged plane under control and was able to land. However, no trace of Peter Sommer was ever found. Carl Spaatz, acting commanding general of the Army Air Forces, wrote about the incident: "My attention has been called to the praiseworthy record maintained by Lieutenant Sommer in the Navigation School at Selman Field, where he completed his navigation cadet training. During his career in the Army Air Forces, he worked tirelessly to perfect his skill and could be depended upon for willing, energetic performance of duty."

For his accomplishments, Sommer was awarded the Air Medal and two oak leaf clusters. His citation read: "The courage, coolness, and skill displayed by this officer upon these occasions reflect great credit upon himself and the armed forces of the United States." He was posthumously promoted to first lieutenant in February 1945, two weeks after his final mission.

James "Bud" McIntosh

James McIntosh was born in Brookline, Massachusetts, in 1922. He was the youngest of three children. His father worked for the Street Railway and the family lived in a third-floor walkup in a wooden tenement located behind the Boston elevated railway car barn. In spite of their modest circumstances, he and his two older sisters never felt deprived. Even with the onset of the Great Depression, the family remained tight-knit and managed to get by financially.

As long as he could remember, McIntosh was fascinated by airplanes. His mother recognized this fascination; often, before putting him to bed, she opened the window of the pantry in their small apartment and allowed him to listen to the drone of the mail plane as it passed overhead en route to New York. After McIntosh entered grade school, he made model planes as a hobby and read every book he could find in the library that dealt with World War I aerial combat. As he later recalled, flying "was an experience which persistently dominated my thinking. I was going to learn to fly."

McIntosh enrolled at Boston University in the fall of 1941. School became far less important to him after Pearl Harbor. He immediately dropped out of school and enlisted in the Army. However, because of the lack of training bases, he was not called to active duty until March 1943. He was then sent to the classification center in Nashville and after that to Selman Field. He completed his Pre-Flight training in June 1943. Following a short break from classroom work at the aerial gunnery school at Tyndall Field, Florida, McIntosh returned to Selman Field for advanced navigation training. After graduating with his wings in December 1943, McIntosh spent a few happy days with his family in Boston. He met his crew in Pueblo, Colorado, where they were assigned to their plane, a B-24. They began to fly practice missions almost immediately. After transferring to Springfield, Massachusetts, they continued the training; it was there he and the men in his crew learned to function as a team. For the final phase of their training, they were transferred to Mitchell Field, Long Island. There they were given intensive gunnery drills as well as practice in bombing, navigation, and formation flying. The end of their training brought new orders; they were being sent to Italy.

The route to Italy turned out to be circuitous and exhausting. Departing Mitchell Field, McIntosh and his crew first flew to Morrison Field, Palm Beach, Florida. After an overnight refueling, they left on April 1, 1944, for Natal, Brazil, with intermediate stops at Trinidad and Belem on the hump of Brazil. From Natal, they crossed the South Atlantic to Dakar, Senegal, where they spent the night while the plane was being refueled and checked out. The next morning they skirted the western edge of the Sahara desert en route to Marrakech; from there they continued on to Tunis, where they were able to enjoy a few days of liberty. After their arrival at their final destination in Cerignola-Stonara, Italy, they were assigned to the 456th Bomb Group, 745th Squadron of the 15th Air Force. They soon received the disturbing news that they would be the replacement crew for one of three planes that had been lost several days before.

McIntosh and his men did not have long to recover from the grueling flight. On April 17, 1944, they were rudely awakened at 5:00 A.M. by a beam of bright light. The men had slept in their winter uniforms for added warmth in the cold, damp tent. They quickly put on their flying coveralls, caps, and leather jackets, and made their way to the mess for breakfast. Then they were hustled onto a weapons carrier for the ride to group headquarters and the briefing for their first combat mission. The group operations officer carefully explained their mission: the target was the railroad marshalling yards at Sofia, Bulgaria. He detailed expected enemy fighter opposition and anti-aircraft emplacements the bomber stream might encounter. He also described the P-51 and P-38 fighter escorts the bombers would have and stressed the significance of the mission. In case cloud cover obscured the target, a secondary target was identified. McIntosh took copious notes throughout the briefing; when it was over, he went to operations to gather mission maps and an individual "escape kit" containing cloth maps and currency of the countries over which he would be flying.

"K" rations were distributed, but McIntosh preferred to carry a few candy bars and a pack of cigarettes. Around his neck were his ever-present dog tags, the Episcopal Church War Cross, and a rabbit's foot for good luck. The crew was then taken by jeep to the runway where their B-24, the "Alberta Kay," was waiting. They put on their Mae West vests and parachute harnesses over their jackets. Each crewmember then went over his final checklist. The gunners tested their twin-.50 caliber machine guns and ammunition; the pilot, co-pilot, and flight engineer visually checked the aircraft and discussed its overall condition with the ground crew; the bombardier checked out the Norden bomb sight and the bomb load, and McIntosh carefully reviewed his maps and flight plan and familiarized himself with the cockpit's navigational instruments. He knew the flight would be long, six-to-eight hours, and had been forewarned of the potential dangers. He hoped he would measure up.

After the crew was cleared for takeoff, the plane hurtled forward and lifted off the runway to join up with the other bomber squadrons flying in a formation of diamond-shaped boxes. Nervous but excited that his first mission had finally arrived, McIntosh awaited the unknown. The gunners, on signal from the pilot, test-fired their guns while McIntosh busied himself checking the ship's location and determining the E.T.A. to their I.P., or initial point. It was there the plane would alter course for the bombing run. The plane crossed the Adriatic; just off the Yugoslavian coast, it leveled off at 18,000 feet.

McIntosh put his twenty-pound FLAK vest on over his parachute harness and placed the unwieldy and heavy FLAK helmet over his cap and headset earphones. At the I.P., the bomb-bay doors opened and, as the bomb run began, anti-aircraft bursts began to cloud the sky around them. The gunners yelled out a warning over the intercom that the fighter escort was heavily engaged with enemy fighters overhead. The difficulty of the mission was compounded by the frigid temperature: -20 degrees Celsius. Though McIntosh was wearing his heated flying suit and gloves, his hands were freezing and because he had chosen not to wear the heavy fleece-lined boots, his feet felt numb. However, in spite of the intense FLAK, the bombing run went well; the ship made a steep diving turn and charted a heading for home. Fortunately, the flight home was uneventful. After a thorough debriefing, the crew was provided with coffee and doughnuts. Then McIntosh and the rest of the crew did what was to become a habit: they tried to unwind in order to prepare for the next mission.

The next few months of combat were intense and emotional. The days seemed endless and the bombing unrelenting. Assigned targets focused on railway marshalling yards, manufacturing plants, and airfields. The men learned during briefing sessions the urgency of taking out the oil refineries and storage facilities at Ploesti, Rumania, in order to cripple the enemy's resources. On May 5, 1944, on a run to Ploesti, the formation ran into intense and accurate FLAK over the target. The bomber just above and to the right of McIntosh, after being hit by FLAK, began trailing smoke. A crewmember bailed out; however, as he fell, he hit the No. 4 engine of a bomber immediately below. His body severed the wing, causing the plane to spiral out of control. McIntosh did not see any chutes from the stricken plane. He was stunned by the tragic accident he had just witnessed. However, constant conflict made it difficult to mourn the dead and or deal with the pain.

On May 10, 1944, the thirty-one-bombers arrived at the I.P. at 18,000 feet and began their bombing run on the aircraft factories at Weiner-Neustadt, Austria. The anti-aircraft fire was relentless but the bombers maintained course and dropped their bombs; then the black bursts of FLAK over the target suddenly turned to pink. Before the crew realized what had happened, the formation was attacked by eighty to ninety German fighters. Only then did McIntosh realize that the pink FLAK was the signal for the fighters to press their attack. McIntosh watched in horror as a nearby bomber was strafed by rocket fire from a German ME 109. The bomber began to lose altitude and exploded at 10,000 feet. Again, McIntosh saw no chutes. At that moment, the "Alberta Kay" was attacked head-on by an ME 109. Cannon fire ripped into the "Alberta Kay's" left wing, tearing a deep gash across the top of the leading edge. The plane shuddered and made a steep turn to veer away from the swirling dogfight. None of the crew was injured, but the plane sustained critical damage. Losing altitude, it managed to clear the 6,000-foot mountain range of the Yugoslavian coast near the Adriatic. Then the No. 4 engine shut down. Fighting to maintain altitude, the pilot was forced to feather the No. 2 engine. Still the plane continued to fall further behind the formation and lose altitude. Soon it had dropped to 200 feet over the Adriatic. Knowing they could not make it to their base, the crew managed to make an emergency landing at an airfield along the Italian coast. The crew was then transported to their base. They arrived that evening just as their tent was about to be cleared of their belongings: the plane had been presumed lost. The "Alberta Kay" never flew again. The crew was later credited with one ME 109 and one probable of the seventeen German fighters shot down. The bomb group earned its first Distinguished Unit Citation for the mission. Of the thirty-one bombers that had departed, five were lost in action that day.

Often, in the short intervals between missions, out of concern for their buddies, McIntosh and his fellow crewmembers would walk to the control tower to count the returning planes as they landed and watch for any signal flares indicating wounded aboard. On May 23, following a hazardous mission, McIntosh and his crew were sitting in the rear of an open truck, preparing to be driven to headquarters for a debriefing. Planes were still circling overhead in a final approach prior to landing. Suddenly, two bombers collided, crashing onto the field. The crash was followed by violent explosions. This tragedy deeply shocked McIntosh, bringing home to him yet once again just how vulnerable they all were.

Typical fare at the mess consisted of Spam, Vienna sausage, powdered eggs and milk, tasteless coffee, white bread, and canned fruit cocktail. Bill Jenkins, McIntosh's bombardier, so disliked Vienna sausage that he came up with a plan. He waited until orders came for a June 16 mission to Vienna. After successfully bombing oil storage facilities on the outskirts of the city, Flight Engineer Boyce Duncan flung a small parachute with a pouch containing a can of Vienna sausage out of the bomb bay. Jenkins attached to the can a note in German that read "Please eat this, we have had enough."

As the war raged on, bomber command devised a plan to mount an all-out attack on the Shell Oil Refinery in Budapest, Hungary. McIntosh's crew took off on this mission on July 2. As they hit the I.P. on time, flying at an elevation of 20,000 feet, the formation began its bomb run. Immediately, concentrated FLAK riddled the planes. A burst of shrapnel shattered the plexiglass of McIntosh's plane, striking his FLAK vest and knocking McIntosh down. At that moment, the anti-aircraft fire stopped and sixty ME 109s and FW 109s swept into the formation. While the fighter escorts fended off the attackers, pilot of McIntosh's plane fought to maintain position. He feathered the No. 3 engine, but the plane began to lose airspeed and altitude, forcing the bomber to drop out of formation and limp home on its own. Knowing German fighters sought out lone aircraft as prime targets, the gunners remained alert; fortunately, no enemy planes were sighted. The raid was deemed a success, but it came at a high cost: six out of thirty bombers were shot down. The bomb squadron was awarded its second Distinguished Unit Citation for that day's run.

With their nerves frayed from the constant tension, McIntosh and his crewmembers looked forward to a few days' rest. But bomber command wanted to maintain the unrelenting pressure on the enemy. So the following day, July 3, McIntosh and his crew were assigned another mission, this time to bomb the locomotive works in Bucharest, Rumania. This was to be McIntosh's twenty-fifth mission, halfway through his sorties. Moderate, but accurate anti-aircraft fire greeted the crew as they began their approach to target, but no German fighter planes appeared. As they left the target area, their relief turned to apprehension. Over the intercom, the bombardier reported in a highly agitated voice that they had a "hung-up bomb." As McIntosh hurried to the open bomb bay, he felt the sting of the frigid wind. He took off his oxygen mask and parachute so that he could squeeze between the bomb racks and stand on the ten-inch-wide catwalk. The flight engineer was already there, trying to forcefully pry the bomb loose from its shackles. Due to the limited space, McIntosh could only watch, wait, and hope. After what seemed an indeterminable amount of time, the bomb was loosened enough so that McIntosh was able to help kick it out.

Following this raid, McIntosh and his fellow crewmembers got a well-deserved rest. They spent the next week on the Isle of Capri soaking up the sun and trying to forget about the war. McIntosh knew that the strain was beginning to tell on him physically. His sleep patterns were disrupted and he suffered a loss of appetite. While the rest served to calm him, it did not seem nearly long enough. On July 22, on the first mission after their return from Capri, McIntosh learned that the target was once again the well-defended Ploesti oil fields. The briefing officer explained how the Germans had mounted 88 and 105-millimeter anti-aircraft guns on mobile railway flatcars on tracks encircling the refineries. When bombers flew over, the Germans would move the flatcars and rotate the guns to direct fire in the direction of the attack. The briefing officer continued, saying that that day's mission was to be one of the largest assaults in the continuing effort to render Ploesti all but useless to the Germans. Four hundred and ninety-five bombers were to be dispatched. The bomber stream took off, accompanied by a fighter escort. The men knew what to expect. In spite of fierce enemy opposition, the attacks were successful. All targets were hit, but at considerable cost: twenty-four planes had been lost. McIntosh was grateful he had lived through another day of combat.

The summer turned to fall and the days became a blur for McIntosh and his fellow crewmembers. As McIntosh neared the "magic fifty" which would complete his tour, he tried not to pay attention to the prevailing superstition that held that the last five missions were the most dangerous. This is when, according to the veterans, you either "got it" or "bought the farm." He noticed that he had developed a tremor in his hands. Like many of his fellow airmen who also had signs of premature aging, he attributed this symptom to severe stress. On October 16, McIntosh was part of a raid on an assembly plant in Steyer, Austria. Anti-aircraft fire was moderate, but no enemy fighter planes appeared. However, on the way back to base, McIntosh saw one plane veer off into a tight spin and crash into the Adriatic. Of the thirty-eight bombers on the raid, this was only one shot down

Shortly after the Steyr raid and a few weeks before his twenty-second birthday, McIntosh was promoted to first lieutenant. He chalked his survival up to pure luck rather than to any accomplishment on his part. Finally, November 7, 1944, the day of his last mission arrived. The destination was Yugoslavia. As had been his custom since the beginning, during the bomb run McIntosh silently recited the 23rd Psalm; this time, though, he focused on the meaning. The prayer helped to calm him and curb his fears. When he had finished praying, he realized that he had not heard any FLAK nor motors of German fighters. With German forces retreating northward in the face of the Soviet onslaught, McIntosh's final mission turned out to be a "milk run," a very uncommon experience.

Following a farewell party by squadron well-wishers, McIntosh left for Naples, where he boarded an ocean liner for New York. At the end of the trans-Atlantic voyage, the troops shouted and cheered as they entered New York harbor. Upon sighting the Statue of Liberty McIntosh was overcome with emotion. Before reassignment, McIntosh took a few days leave to marry his long-time sweetheart. He then reported to the navigation-training base at Ellington Field in Houston, where he served till the end of the war. He was discharged in October 1945.

EDWARD M. SIANTZ

Edward Siantz arrived at Selman Field from Detroit on November 13, 1943, to begin the rigors of advanced navigation training. He quickly settled into military life, with both day and evening classes. Reveille was at 05:30 and lights out at 22:00; in between he had little spare time to himself. The unusually cold temperatures that winter made life even more uncomfortable in the damp, uninsulated barracks. The one thing he looked forward to was mail from home. Whenever he did not receive any letters, he anxiously hoped for mail the following day. Sometimes, in the few evenings when there were no classes, he would take a break by seeing a camp movie or a traveling USO show.

Thanksgiving was an especially difficult time for Siantz. This was his first time away from his family and he was homesick. He felt somewhat better when he got the results of his daily and weekly tests, which usually placed him in the top quadrant of his class. Also, his first payroll check for $45, which he received just after Thanksgiving, lifted his spirits. He also began to bond with those in his class. Finally, he began to receive his mail: eleven letters on November 22 and fifteen on November 23. As he later indicated, "I was a happy cadet." Feeling upbeat, he went into town to treat himself to dinner and a show with his new paycheck.

By the beginning of December, he was ready for the twenty-one missions he needed to graduate. He was excited at the prospect; however, he became airsick on his very first flight. Afterwards he was miserable, knowing he had failed his first major test. He was determined not to let that happen again.

Christmas day was especially hard for him. Even though there was a traditional Christmas dinner, he still missed his family. He waited nine hours to call home, without any success.

Over the next three months, Siantz flew his required twenty-one missions. Each flight added to his self-confidence, as he became more comfortable with the complexities of navigation. His training flights took him all over the Southeast: to Florida, Georgia, Mississippi, Arkansas, Oklahoma, and Texas. By the time he entered his twelfth week of training, he was sure he was going to earn his wings. Indeed, during that week he was instructed to place an order for his uniform. He was thrilled until the officer at the PX told him it would cost $168.50, a price he considered exorbitant.

Finally, on March 18, 1944, in ceremonies at the post theater, Siantz received his wings as well as promotion to second lieutenant. His parents came to Monroe to attend the ceremony and he gave his mother the honor of pinning on his wings. It was an emotional day for Siantz, but it was fleeting. That night he was on a train bound for Westover Field, Massachusetts. There, Siantz was assigned to the 15th Air Force. By war's end, he had safely completed thirty-five missions in his B-24. He flew his last mission on December 15, 1944, and was married in February 1945.

Edward Siantz passed away on June 10, 2000. He had been married for fifty-five years and left three children and four grandchildren.

Joseph Rifkin

Joseph Rifkin completed navigation training at Selman Field in May 1944. However, he was given the lesser rank of flight officer rather than second lieutenant. Rifkin had received word of the reduction in rank via the base "grapevine" prior to graduation. He had previously had disagreements with one of his instructors and was therefore not surprised by the demotion. He said later: "no one, regardless of rank, can tell me day is night and night is day and expect me to agree." Yet, the strong-willed Rifkin felt he would make a good officer. Although he did not receive his wings early on, his goal was to be the best navigator possible and get into combat. His ratings later reflected his self-confidence and were never lower than excellent.

Rifkin joined his new crew and became familiar with them during training. Simulated bombing missions helped the crew to function as a close-knit unit. They were assigned to Bari, Italy, where the 345th Bomb Squadron, 98th Bomb Group, 47th Wing, 15th Air Force was headquartered. Rifkin was thrilled to be assigned to the 98th Bomb Group. Its reputation had spread throughout the Air Force. Officially activated in February 1942, at McDill Field in Tampa, the 98th had undergone intensive training before being shipped to Haifa, Palestine. There it provided direct air support for General Montgomery's British 8th Army against the Afrika Corps, commanded by Gen. Erwin Rommel, the "Desert Fox." With Rommel's withdrawal across North Africa before the victorious British, the 98th Bomb Group continued to provide strategic air support. As the allies surged forward, the 98th moved into new forward bases just behind the front, eventually winding up in southern Italy. The actions of the 98th Bomb Group earned it two Presidential Unit Citations for its precision bombing and the devastating effect it had on the enemy. Rifkin's crew was ordered to Hampton Roads, Virginia. There, on September 9, 1944, Rifkin and 350 other B-24 crew members comprising thirty-five B-24 crews, boarded the Liberty Ship, *Robin E. Lee*, as part of a large convoy bound for Bari, Italy. Slipping out of port at night to avoid German U-boats, the ships began their hazardous journey guarded by escorts, destroyers, and an aircraft carrier. Living conditions deteriorated quickly. The men were crammed into the hold where metal cots were stacked up five high. The heat and foul air below decks made sleep almost impossible. There was only one small bathroom, five by seven feet, for all the men. The only available water was salt water and that had to be used for showering, shaving, and laundry. On the second day out, a huge storm smashed into the convoy, knocking out the ship's refrigeration system. All foodstuffs had to be thrown overboard and only "C" rations were available for the rest of the trip. The days passed slowly. To occupy their time, the men took long walks around deck, played cards, wrote letters home, read books, and served as unofficial lookouts for submarines. Finally, on October 9, the ship entered Bari. Of the 350 airmen aboard, twenty percent became casualties within the next six weeks.

After disembarking, Rifkin's crew was processed; the men arrived at their designated field located near Lecce, Italy, on October 11. Normally new crews did not fly the first mission together. In order to get a feel for combat in the company of more experienced airmen, the crews were split up so they could fly with veterans who had already flown at least five missions. Rifkin was introduced to the crew of the "Queen of Hearts." But he hardly had time to become acquainted with his new crew before leaving on his first mission five days later. The day's target was the Herman Goering Tank Works in Steyer, Austria. Rifkin attended his first briefing, and as was to be his custom, he meticulously worked out a flight plan, calculating his E.T.A. based on all available information. Only twice in combat was he ever off by more than one minute of his estimated time. Unfortunately, this was to be one of those missions.

When Rifkin arrived at the "Queen of Hearts," the crew chief was waiting for him. He wished Rifkin good luck; Rifkin, not knowing what to expect, replied that he "would probably need it." He did not realize at the time right he would prove to be. Although the trip to the target was fairly uneventful with the bomber formation encountering only moderate FLAK, the trip home more than made up for that. The pilot radioed Rifkin, urgently asking for an immediate heading to the nearest airfield. The plane, he said, was quickly running out of fuel. Rifkin figured that the nearest friendly field was twenty miles away. The crew prayed their fuel would last. A few minutes later, the plane dropped rapidly in altitude until it was little more than 500 feet above the vineyards below. The pilot alerted the crew to prepare for a crash landing. Rifkin later recounted the crash: "My bombardier Bill Adams was not wearing his head set, so he had no way of hearing the order. Since we were both in the nose of the plane, I grabbed him by the shoulders, lifted him off the floor, and shouted into his ear what was happening. We wasted no time in scrambling under the pilot's deck through the bomb bay and managed to get into the rear of the plane while it was going down. In fact, while moving through the bomb bay, we felt the pilot practically stand the plane on its right wing as he turned the plane so as to take advantage of the uninviting terrain. Bill and I reached the rear of the plane. I propped myself against the bulkhead and put him in between my legs. What was racing through my mind during the fleeting seconds was the fact that I was mad and saying to myself, 'Just like me to get killed on the first mission.'"

Rifkin blacked out for a short time; when he regained consciousness, he found himself with his feet on top of the rear deck while the rest of his body faced downward against the bulkhead. His first thought was to look for Adams. Finding him pinned under the bull turret, Rifkin tried to lift the bull turret off Adams' leg. Unsuccessful, he asked the other crew members for help; instead, fearing the plane would burst into flames at any minute, they ordered him to evacuate with them. Rifkin refused, but they forced him to leave at gunpoint. However, when it became evident the plane was not about to explode, the crew hurried back to the plane. Two of the crew were trapped in the plane, Bill Adams and the top turret gunner, everyone else seemed all right. The men, working together, were able to free Bill Adams, though he was in terrible pain; unfortunately, his leg was crushed, thus putting an end to his tour of duty before it really began.

Within fifteen minutes, friendly Italians arrived to assist the survivors. They offered the Americans wine and one of the Italians, who had a horse, offered to take one of the crewmen to the nearest airfield, nine miles away, to get help. The crew had to wait for the men to return with the necessary tools so they could free the gunner. Five hours passed while Adams waited in pain from the serious leg injury he had sustained. As a result, Rifkin and the rest of the crew did not make it back to base until late that evening. It was not until then that Rifkin realized he also had been injured. For the next three weeks, he received medical treatment for a severely damaged nerve in his left leg which has caused him pain and stiffness ever since. When he was able to fly again, he superstitiously worried if anyone wished him "good luck" before take-off.

After his eighth mission, Rifkin volunteered for extra-hazardous duty on combat missions as a radar-navigator-bombardier. This would assure him of promotion to second lieutenant as well as supplemental pay. This mission would also hone his skills as a navigator. However, on the downside, it meant that in the meantime he could not fly with his original crew. Several times after his initial crew returned from a mission with a replacement navigator, they accused Rifkin of "trying to get them killed," since he was no longer there to plot their course.

As his flights piled up, Rifkin became a seasoned navigator and bombardier. Briefing information for missions over Vienna included escape and evasion plans in case the plane was shot down. The crews were told that if they managed to parachute to safety, they should take Trolley 132 to a location where they would be picked up by partisans. Trolley 132 could be recognized by a circle with a dot in the center and a line through the dot. The crewmen never quite understood how they were expected to land inconspicuously in a densely populated area, hide their chutes, and blend in with the local population, all the while wearing their flight suits. Nor did any of them speak German. Fortunately, Rifkin never had to figure this out.

Once in a while there was even cause for laughter. On one bomb run northeast of Vienna at the Schwechat Oil Refinery, the B-24 was flying straight and level at 24,500 feet as it approached the target. Suddenly the plane was bracketed by three shell bursts. The first shell exploded under the left wing, causing upward pressure that forced the plane to veer to the right. The second shell exploded beneath the right wing and this caused the plane to veer to the left. The third shell exploded under the open bomb-bay doors and blew the plane upwards until it reached the proper altitude for release of the bombs. En route home, the crew tuned in to Radio Foggia to listen to some relaxing music. One song featured Lionel Hampton singing "Straighten Up an Fly Right;" the whole crew broke out in laughter. "How true it was," Rifkin thought, as they ended another stress-filled nine-hour workday.

On February 25, 1945, Rifkin was part of the massive 1,000-plane 15th Air Force raid on the heavily defended railroad marshalling yards at Linz, Austria. FLAK over the target was intense and accurate. Rifkin decided to use the radar sight to bomb the target rather than try to bomb visually. Reconnaissance photos later showed that the seven planes of the 345th Bomb Squadron for which Rifkin was responsible accounted for over ten percent of the damage on the raid: one hospital train, forty-seven railroad cars, one railroad station, and substantial railway track were destroyed.

A little over a month later, on March 30, 1945, Rifkin was informed that he would be returning to Linz, this time as a member of the lead crew. He reported to bomb group headquarters that evening to familiarize himself with photos of the next day's target as well as with the radar bombing maps. The next morning, Rifkin completed his flight plan prior to takeoff. This review had become routine. After his near-fatal first mission, Rifkin had made it a practice to review his maps, charts, and recon photos thoroughly each and every time. Perhaps, as he recalled, it was because he was an "untried rookie" that the "Queen of Hearts" had crashed. If he had anything to with it, he would never allow that to happen again.

The bomber stream was forewarned of heavy cloud cover that obscured the ground targets and made radar-bombing a necessity. Once the I.P., the initial point, was reached, the bombing run began at an altitude of 25,000 feet. The planes flew straight and level and took no evasive action. Everyone knew they were "clay pigeons" for the German anti-aircraft gunners. Intense and heavy FLAK and air turbulence which caused Rifkin's plane to be late over the target. Nevertheless, the run was successful, and on the return trip home the crew could finally unwind. They tried unsuccessfully to eat their box lunches, consisting of "K" rations; but with the temperature gauge indicating –46 degrees, the meals had frozen solid. The plane arrived late. After being debriefed by the squadron intelligence officer, Rifkin returned to his tent; there he found several of his colleagues, who had thought that his plane had been shot down, in the process of dividing up his possessions.

It was not until April 15, 1945, that Rifkin rejoined his original crew on what would be their last mission together. The day's target was a bridge in northern Italy. The mission was rather uneventful until the plane was right over the target. Four FLAK bursts exploded around the plane. The last burst blew a hole in the right window of the navigator's compartment. Rifkin happened to be peering out the left window and remained unharmed.

Although most of the 98th Bomb Group was ordered home in April 1945, Rifkin did not receive his orders until June. Finally on June 28, his B-24 lifted off for Marrakech with fifteen men aboard. On the first leg of his return, Rifkin discovered that the radio compass was oscillating; this prevented him from getting a true bearing from any radio beacon. Then the airmen encountered a sandstorm over the Sahara. They decided to make an unplanned landing at Casablanca to check their gas when they discovered their fuel gauges were not working either. They spent the night in Casablanca in order to have the radio compass and gas gauges repaired. Then they flew on to Marrakech where they had originally planned to spend the night. After being assured that the problems were fixed, they took off for the Azore Islands. Over the Atlantic, Rifkin switched on the radio compass only to discover that it had not been repaired. Refusing to panic, he took sextant readings of the sun and estimated the wind direction and speed. He plotted a course and gave the pilot directions, constantly updating his calculations. If Rifkin had encountered solid clouds, had not been able to take sun shots, or if the sextant had malfunctioned, the plane might never have made it to the Azores. Fortunately it arrived safely. After a night of welcome rest, the crew flew on to Gander, Newfoundland, and then to Westover, Massachusetts. There the plane was finally delivered to the 376th Bomb Group.

JAMES INKS

James Inks was born in Llano, Texas, on November 9, 1921. His father, who had served as the mayor of the town, died in 1935, when James was fourteen since his mother had deep roots in Texas, she moved the family to Austin. Her grandfather had fought in the Texas war for independence. Following the Battle of San Jacinto, which secured the state's independence, he had received a large land holding. After graduating from high school, Inks enrolled in the New Mexico Military Institute.

Inks graduated from that school and then entered the University of Texas at Austin, majoring in Geology. But Pearl Harbor prompted him to sign up for the Air Force. Inks was sent to Selman Field, where he graduated as a navigator in 1943. He was assigned to the 464th Bomb Group in Pocatello, Idaho. After training missions were completed, the group flew their B-24s to Tunis in North Africa. Later they were assigned to Italy, as allied forces moved up the peninsula.

On his 43rd combat mission, while on a bombing raid over the oil installations at Ploesti, Rumania, Inks' plane was hit by FLAK. The pilot tried to return to base, but the plane continued to lose power and altitude. Finally, Inks and the remaining crew members had to parachute out over German-occupied Yugoslavia. There, they were captured and detained by royalist "Chetnik" partisans. Inks was in constant pain from wounds caused by shrapnel and aggravated by a rough landing. The men had insufficient food and inadequate clothing to protect them from the bitter cold. They were shuttled from place to place to avoid German patrols. Gradually, though, they began to fit in and were accepted as partisans by their captors. They helped wage guerilla war on two fronts: one against the communists under Marshall Tito and the other against the Germans. After almost eleven months as partisans, Inks and his crew were finally allowed to seek refuge behind American lines. Inks later wrote an account of his harrowing experiences in *Eight Bailed Out,* which proved an instant success.

After the war, Inks went through pilot training; during the Korean War, he served as a troop carrier pilot in the airborne operations at the Inchon Invasion and in the Marine evacuation from the Chosin Reservoir. He retired from the Air Force in 1962 as a lieutenant colonel. His service record reflected forty-three combat operations in World War II and ninety-two combat missions in Korea. He returned home to the family ranch in Llano, Texas, where he lived until his death in 2004.

David Purner

David Purner enlisted in the Army Air Corps in April 1942. He wanted to follow in the footsteps of his older brother, who had enlisted in the Royal Canadian Air Force before the United Sates entered the war and had won his wings as a navigator. In December 1942, Purner arrived in Monroe to begin the pre-flight program. Upon completion, he entered the advanced navigation school at Selman, graduating in July 1943 as a second lieutenant. His first assignment was to Davis-Monthon Field in Tucson, Arizona, there he met the rest of his crew and began daily training exercises in a B-24. After several days, the crew moved on to Blythe Field, California, for additional phased-in training.

In November 1943, the crew was ordered to Topeka, Kansas, to pick up their B-24. During his embarkation leave in Topeka, Purner married his childhood sweetheart; the honeymoon was necessarily brief. On December 21, Purner and his crew took off for their final destination, half a world away. After stopovers in West Palm Beach, Florida, Puerto Rico, Guyana, Belem, and Fortaleza in Brazil, the crew put down in Natal, Brazil, on the northern hump of the country on January 8, 1944. By that time the crew had noticed several problems with the plane: the nose wheel had not been responding well, the main landing gear was not functioning properly, and the wing fuel tanks were leaking. Mechanics worked on the plane for three weeks while Purner and his crew killed time by enjoying the beaches and touring the old colonial city. Once repairs were complete, they headed across the Atlantic to Dakar and from there, north to Marrakech in French Morocco. After a week of rest in Marrakech, they began the last leg of their flight to Britain, flying along the Portuguese coast. This was the most dangerous part of the flight and, the crew, manning the guns, was on full alert. The Germans were known to send out false radio signals to guide single planes into German-occupied territory so they could shoot them down. This flight, however, was without incident.

Purner and his crew reported to the 392nd Bomb Group, 576th Bomb Squadron in East Anglia. Purner had just turned twenty-four, but there seemed little reason to celebrate. Morale was deteriorating. His crew was one of nineteen replacement crews that were to help make up for the high losses of the 392nd. A week after their arrival, on April 1, 1944, Purner and the other crewmembers were ordered out on their first mission to Brunswick, Germany. The airmen were not really superstitious, but they did appreciate the irony of making their first bombing run on April Fool's Day. This proved to be an exceedingly rough baptism by fire, but fortunately they returned safely. Other equally stressful missions followed in quick succession: Zwickau, Augsburg, Tutow, Schwienfurt, and Hamm. Then, Purner and his crew were told they were to be a part of the largest raid of the war.

On April 29, 1944, over 2,000 bombers from the 8th and 15th Air Forces hit targets throughout Berlin in a meticulously planned attack. However, the daylight raid proved to be a costly venture. German fighters and accurate FLAK bursts brought down sixty bombers and fourteen fighter planes. Purner's bomb group alone, the 392nd, lost eight bombers. As Purner's plane flew over Hanover on the return flight, German fighter planes made two passes. Machine gun fire shredded the aircraft, caused numerous fires, ripped out two engines, and decimated the tail assembly section. Purner was wounded in his left leg and foot.

The plane fell out of formation and began losing altitude. Then, below cloud cover at roughly 1,000 feet, it spiraled into a tight spin. Purner could not keep his balance, the plane's rapid descent throwing him against the nose wheel. The pilot had previously dropped the landing gear, making bailout extremely hazardous. The powerful slipstream ripped off Purner's fur-lined boots. Pinned by the centrifugal force against the side of the aircraft, Purner could not move. He thought the end had come as he watched the ground move closer and closer. At about 500 feet, there was a terrible explosion. The airplane broke apart, and Purner was free. He searched for the ripcord on his chest pack with both hands, but when he could not locate it, panic set in. He could not believe his eyes when he looked up and saw that the chute had opened. Smoke and debris were everywhere, obscuring the sky. Before he knew it, he hit the ground hard. He unhitched the parachute and buried it in the field where he had landed. Dragging his injured leg, Purner tried to hurry into the woods. He heard German soldiers searching for him; falling flat on the ground, he remained very still. After the Germans left, he was relieved to see another member of his crew. They hid together, watching the last planes of their formation disappear into the clouds and head back to England and safety. They remained still the rest of the day, moving only at night. Purner's leg wounds were bleeding profusely and a growing cause for concern. He put chlorine pills in water to wash and disinfect the wounds, using the white lining from the cuffs and collar of the shirt of his uniform as a bandage. But walking was painfully slow. Having lost his boots, he had only his socks to protect his feet from the rough terrain.

On the third day, Purner and his companion were captured near Hamburg by angry civilians armed with clubs, pitchforks, guns, and accompanied by snarling dogs. The men were beaten, spit upon, and threatened with hanging. The police took them away for interrogation and then threw them into solitary confinement. Afterwards, Purner was separated from the other crewman; he

Aerial view of Selman Field.

Training flight over Monroe, LA.

Parade on DeSiard Street in Monroe, LA.

Colonel John W. Egan, Selman Field's third commanding officer, from 1943-1944.

Selman Field flight line.

Selman Field obstacle course.

Selman Field obstacle course.

Selman Field mess hall.

In the classroom.

Training plane cockpit.

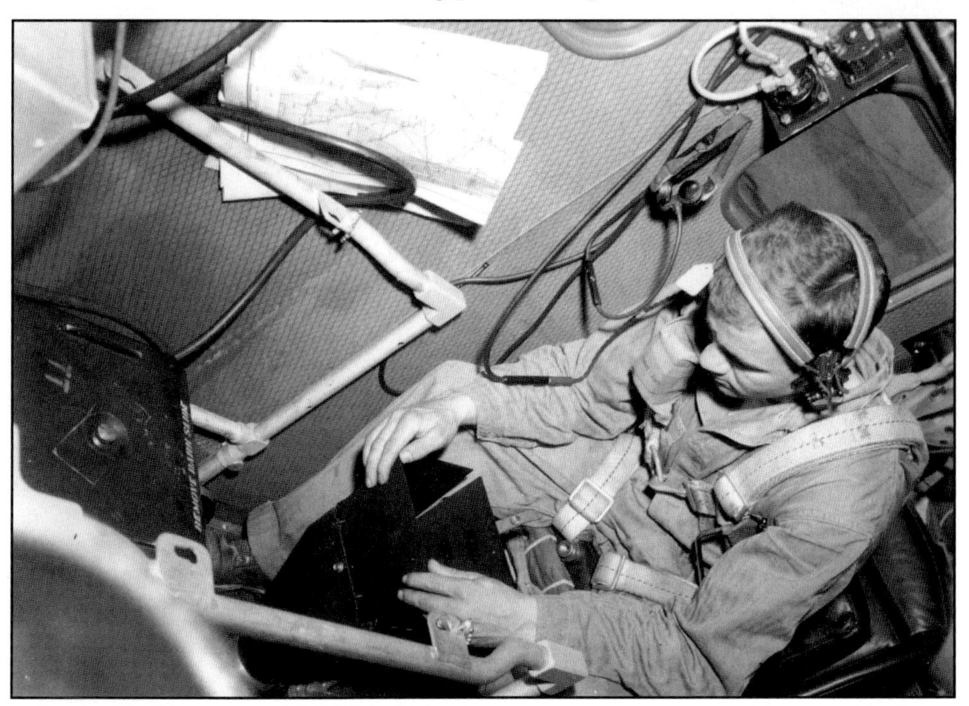

Cockpit training.

Navigational instrument training.

Flightline maintenance.

Selman Field hangar.

Charting the flight.

Entertainment – a regular part of Selman Field life.

One of many USO dances.

USO chaperones.

Selman Field Officers Club.

On parade.

Early construction at Selman Field.

Base bar-be-que.

End result of a very long day.

was thrown into an overcrowded boxcar for transport to Stalag Luft III, about seven miles southeast of Berlin. He never saw any of his fellow crewmen again.

Purner spent the last year of the war in prison. Initially he was placed in the prison dispensary; it was staffed by captured doctors and medics who were forced to make do with inadequate supplies from the Red Cross. By now, Purner's foot was swollen, discolored, and badly in need of treatment. The staff could do little. Purner stayed off his feet, kept his wounded foot elevated and applied saline compresses. After seven weeks, his wounds had healed sufficiently so that he had regained his mobility and was pain free. Purner was assigned to a cooking unit even though there was little to cook, given the meager food allotments provided by the Germans. Only the inadequate rations from Red Cross parcels stood between the men and starvation. The food was carefully distributed at the rate of one parcel per eight men each week. By mixing some of the food with water to make a watery soup or stew, the men were able to make the food last longer. But even so, the food did not last through the week and the men suffered constantly from hunger pangs.

Shortly before Purner's arrival, a mass escape attempt made famous by the film *The Great Escape,* had been thwarted. The Germans reacted ruthlessly, inaugurating a reign of retribution. They randomly selected forty-seven airmen for execution. Then, soon after Purner was assigned his cooking duties, the Germans quit distributing Red Cross parcels altogether. The inmates were reduced to eating grass and making soup from horse bones. Nor did the Germans distribute any clothing or blankets during the bitterly cold winter of 1944-45. Furthermore, they did not permit the men to cut firewood to heat their barracks; the prisoners were forced to huddle together to provide themselves with warmth. No soap was distributed and only one cold-water spigot served all 350 prisoners for washing and drinking. Vermin and rodents infested the barracks, and lice ate at the men's ankles, wrists, and hips, causing constant itching. The toilet facility was a G.I. can with a board across the top, which overflowed each night. Not surprisingly, dysentery, jaundice, and typhus began to spread throughout the prison.

In January 1945, with Russian forces beginning to encircle Berlin, the prisoners were force-marched to Nuremberg. Snow and ice covered the ground. The temperature was fifteen degrees below zero and no food or water was supplied during the first three days of the march. No rest periods were permitted and those who could not keep up the pace were clubbed to death or shot. Bodies littered the road. Purner survived by drinking melted snow and eating dried prunes, lumps of sugar, and one "D" ration bar he had hoarded from the

Red Cross parcels for just such an emergency. When the survivors arrived in Nuremberg, their feet and hands were frozen, and most of the men, including Purner, suffered from frostbite. They were thrown into a barbed wire encampment next to the rail yards. Though the rail yards were under a fifteen-day air assault by American and British bombers, the sight of allied aircraft gave the men reason to hope and raised their spirits.

The prisoners remained at Nuremberg for two months. They were then force-marched to a second camp that was just as filthy and unsanitary as the first. Food and water were almost nonexistent and there was no medical care. In March 1945, with American forces fast approaching, the prisoners were force-marched once more, this time south toward Munich. Prisoners slept in the open air on the cold ground at night. They were forced to forage for themselves for whatever provisions they could find as they shuffled through the towns. On three different occasions, the prisoner column was strafed by Allied planes, killing and wounding many. When the surviving prisoners finally arrived at Moosburg, just outside of Munich, they were more dead than alive. They were funneled into an encampment with over 100,000 Allied prisoners of war. Purner weighed only ninety-five pounds and did not think he could have walked even one more mile. By now, German organization and discipline had broken down. Little food or water was supplied. Sanitary facilities were nonexistent. The men lived outside, lying on the ground, waiting to die. Finally, on April 29, 1945, after a brief but fierce firefight, American troops overran the German defenses and liberated the camp. David Purner had survived.

William Holbrook

William Holbrook was born in 1921 in Akron, Ohio. Following graduation from high school, he entered the University of Akron. However, after one year, he left to work as a draftsman at the Goodyear Tire Company. In April 1942, four months after Pearl Harbor, he quit his job and enlisted in the Army. Four months later he got married. Holbrook was then sent to the classification center at Nashville, where he was selected for pilot instruction and dispatched to Marfa, Texas, for training. He received his silver wings and shortly thereafter was ordered to report to Selman Field to work on a "P-38 Project." Eight men were selected for the program, four of whom had volunteered and four to replace volunteers who had not earned satisfactory scores in navigation training. The adjutant to whom Holbrook reported had no knowledge of the mission; his only instructions were to tell the volunteers that they were not allowed to fly. They were there to take an intensive navigation course with emphasis on celestial navigation. After a month of accelerated classroom work, Holbrook and the others were ordered to Pensacola Naval Air Station for further instruction.

The eight men began training in a PBY, a large, twin-engine, ungainly, and slow- moving amphibious flying boat. Since the Army had no such facilities, the Navy assumed responsibility for the training. Holbrook received his Navy wings in January 1944. He was assigned to the Second Emergency Rescue Squadron of the 13th Air Force. His job was to provide air-sea search and rescue for aircraft on combat missions in the southwest Pacific. The Squadron consisted of four flights. Each flight included three Catalina flying boats and operated as an independent unit, with each flying boat manned by a ten-man crew. Holbrook was ordered to serve as the co-pilot to the operations officer, the second in command of the squadron.

Holbrook's trip to the combat zone was long and tiring. After completing a long cross-country flight to Sacramento, California, the crew flew 19½ hours, first to Hawaii, then to Townsville, Australia, with intermediate stops at Canton Island, Tarawa, and Guadalcanal, scenes of fierce fighting earlier in the war. At Townsville, where the principal maintenance facility for the 13th Air Force was located, the plane underwent a thorough inspection; it was repainted a dull blue color so it would blend in with the sea. When the plane was ready, the crew flew to New Guinea and then to their combat base on Biak Island. There the crew was assigned to hastily-erected tents atop the dirty coral. Holbrook and his men were given little time to settle in. They were soon made aware of just how needed they were. Because of a shortage of planes, the crews operating out of Biak had to fly to the point of exhaustion. Bombers and fighter planes were flying maximum range missions, which took a minimum of twelve hours to complete. If there was a search for downed airmen, a mission took fourteen to eighteen hours. Holbrook flew 100 hours the first month and 135 hours the second month he was at Biak.

Holbrook learned quickly about the hazards of open-sea landings for both the plane and the crew. The hulls always sustained damage in an open-sea rescue. Often the plane's skin buckled and rivets popped in the navigator's compartment. Crews too suffered from the constant strain of long over-water flights; the stress was intensified by the possibility of the defenseless ship encountering enemy fighter planes. Nevertheless, the rugged dependability of the PBY saved the lives of Holbrook and his crew on numerous occasions. During the year in which Holbrook flew rescue missions, the squadron lost all of its original planes. Four were destroyed on the ground in enemy bombing raids and the rest in difficult amphibious landing operations. On the plus side, only five members of Holbrook's squadron were lost while over 700 airmen were rescued.

Charles Titkemeyer

Charles Titkemeyer was born in Indiana in 1919. He was the son of a German-born father who had come to this country at the age of fourteen with his parents. His father worked at numerous odd jobs, learned English, and finally saved enough money by working as a farmhand to purchase his own land in rural Indiana. In his youth, Titkemeyer worked hard on the family farm; in spite of the depression, the family did not want, for the farm produced an abundance of crops.

After graduation from high school, Titkemeyer worked for three years before entering the Purdue University School of Agriculture. In January 1942, a month after Pearl Harbor, he volunteered for the Army Air Corps, hoping to be accepted into pilot training. He began the rigorous program, but after eight weeks, he was directed into navigation training. After completing the pre-flight navigation program at Selman Field in August 1942, he started the advanced navigation school in October. Titkemeyer found the training extremely challenging. As he later recalled, "It required a tremendous knowledge of mathematics and an ability to make a quick decision. I soon learned that the Air Corps was very pilot-oriented and that the pilot always got the credit when a crew was successful. I also soon realized that the pilot was a glorified truck driver and that the real success of the mission depended largely on the ability of the navigator."

After graduation, Titkemeyer reported to Tuscon, where he and his new crew began the process of getting to know one another as well as their new bomber. Gradually they became a smoothly functioning team. By the time six months of training missions in Tucson, Arizona, and Clovis, New Mexico, had elapsed, the members of the crew had confidence in each other and felt well-prepared; they were eager for combat.

In June 1943, they received orders to fly to North Africa. They first flew to Bangor, Maine, and then on to Gander, Newfoundland. From Gander, they left for England. Halfway across the Atlantic, they were ordered back to Gander due to turbulent weather. They returned, bitterly disappointed. Their second attempt at the trans-Atlantic flight was almost aborted as well. At the pre-flight weather briefing, the planned flight was cancelled due to lack of a tailwind. However, Titkemeyer's crew requested permission to depart since they had a B-24 and not a smaller B-17. Permission was granted, and they were finally on their way. Cloud cover obscured the ocean during the flight so Titkemeyer had to check his course by celestial navigation; but since it was dark for about three hours of the day, he could only get in two quick fixes. Fortunately, this proved sufficient to get the plane to base. From England, the plane flew to Oran in French Morocco and then the crew flew the last leg, over eight-and-a-half hours, across the Sahara, to Benina Airfield in Libya. There they landed safely on the dirt runway.

During the next few days, Titkemeyer and his crew familiarized themselves with the field. Debris from the tremendous fighting between Montgomery's 8th British Army and Rommel's Afrika Corps had not yet been cleared away. Over 200 burned German and Italian planes had been bulldozed to the sides of the runways. Most of the buildings on the field had been destroyed by the fierce fighting and those that remained were riddled with bullets.

Following Titkemeyer's arrival, he and his men were given an orientation dealing with living and flying conditions. They then drew equipment, including heated flying suits and flying boots, from the quartermaster sergeant. Four days after his arrival, Titkemeyer and his crew were given their first assignment—the Ploesti oil refineries, cracking plants, and distillation units. The men learned during thorough briefings that the refineries produced over one-third of all the fuel used by the Axis. A successful attack would mean an early end to the war, which could possibly save 200,000 Allied lives. The attack on the heavily defended area, however, had to be a complete surprise; if not, it would be a suicide mission. The briefing officer stressed one more thing: once the bombers turned on the bomb run from their initial point, or I.P., to the target, no evasive action was to be taken. All planes were to fly straight and level to ensure a successful mission.

The attack would employ a new tactic: a low-level run by heavy bombers. This had never been tried before and Army Intelligence believed it would catch the enemy by surprise. Previous attacks employing high-level pattern bombing had been hit or miss and were not always accurate or effective. Thus, the decision was made to employ a concentrated and aggressive attack—it was thought this would have a greater chance of success. The men left the briefing quiet and subdued, each one immersed in his own thoughts. This was the first time it occurred to the airmen that they might be expendable.

Titkemeyer spent the next day studying the target area on the maps. He carefully plotted the course to the target over 2,400 miles away, memorized all important details, and drew his maps. He also viewed a film by Army Intelligence which showed all the targets of the raid. In the meantime, the rest of the crew cleaned the ship's guns and readied the plane. On August 1, 1943, Titkemeyer and his crew were awakened at 4:30 A.M. The airmen ate a quick breakfast, even though no one had much of an appetite.

Within minutes of their departure, 175 bombers had lined up in formation. The weather was fair and clear; vapor trails crossed the sky. The long trip across the Mediterranean was quiet, each man contemplating the fate that awaited him that day. The formation gradually turned for the flight across Greece, Yugoslavia, and into Romanian air space.

Titkemeyer described the next few minutes as "flying into the jaws of death." After making its turn at the I.P., the plane began its bombing run. The plane zipped forward at a speed of 220 miles per hour, all the while hugging the ground so closely that it clipped the tassels of the corn in the fields. Titkemeyer noticed an extended row of anti-aircraft batteries blasting away at the incoming planes. FLAK burst about ten feet above the plane and continued to follow it all the way to the target. The top turret, waist, and tail turret guns aboard the plane opened up on the batteries with withering and effective fire. Titkemeyer saw one of the German gunners topple off his platform. Then he heard an explosion as the plane in front of them went down in flames. Almost immediately, the plane behind them went into a dive and smashed into the forest. Then the plane behind that one was also hit; it shuddered, went into a spin, and exploded as it hit the ground. He saw no parachutes from any of the three planes. Titkemeyer was stunned. Great clouds of smoke were everywhere. His plane rose slightly to clear the buildings at the target, dropped its bomb load, and then dropped to just above ground level, speeding off at 240 miles per hour. Titkemeyer was able to make out the town through breaks in the thick smoke and flames. Thousands of armed and angry people stood in the streets firing revolvers, machine guns, and rifles at the planes. Bullets were everywhere. Just then, a plane following behind them hurtled head on into a tall smoke stack; the result was an enormous explosion.

As the American bombers fled the hailstorm of lead and steel, German fighter planes swooped down on them, triggering an enormous firefight. During the melee, the waist gunners pitched several incendiary bombs into the woods less than 200 feet below; Titkemeyer and the other crewmen saw tank cars, storage tanks, and gun emplacements hidden in the trees, just before the forest burst into flames. As Titkemeyer's plane streaked away, the German fighters did not pursue them. The airmen had their first chance to assess the situation: the plane had weathered the storm. The only damage was to the radio antenna, which had been torn off by FLAK. The crew also noticed a small dent in the wing caused by a tree during the time the plane was flying at treetop level.

The trip home was uneventful. The thirteen-and-a-half hour flight back to base gave Titkemeyer time to reflect. The baptism by fire on his first mission was one he would never forget. Nor would he ever forget the terror he felt or the airmen who died. He was drained of emotion and thankful to be alive. When the plane arrived back at the base, after the crew was debriefed by Army Intelligence, the men learned that the raid was considered a success; most of the targets had been destroyed and fifty-six German fighter planes had been shot down. But the severe losses gave the airmen pause for reflection. Of the 175 planes that had taken part in the raid, sixty-six had been shot down, most of them over the target; in addition almost two dozen were brought down en route home. In Titkemeyer's squadron, only four of the nine planes made it home safely. Search planes were dispatched to scour the Mediterranean for those planes that might have run out of fuel and ditched in the sea; fortunately, several crews were located. The next day memorial services were conducted for the hundreds of men killed on the mission. The knowledge that so many of their comrades were gone had a devastating impact on the survivors. They were grateful to be alive; at the same time they mourned the friends they had lost and knew they would be lucky to survive the days ahead.

Two days later, on August 3, the colonel called all the men who had participated in the raid to a special briefing. He informed them that never before had such a large number of heavy bombers participated in such a low-level assault over that far a distance: "You men who have returned from this trip can now say you have seen Hell, have rode through it, and have lived to laugh about it afterwards." The men were congratulated and called heroes. The British Broadcasting Company referred to the raid as "unparalleled in the annals of heavy bombardment. It was the most daring, most vital, and most destructive raid ever pulled." Then the men were told to put the raid behind them and prepare for the next mission. Titkemeyer thought nothing could be worse than what he had just experienced. He recalled an old saying: "Today is the tomorrow you worried about yesterday."

Titkemeyer's second mission was scheduled for Friday the thirteenth. Although Titkemeyer was not really a superstitious man, he still would have preferred another date. The briefing officer informed the crew they were to hit a Messerschmidt airplane factory at Weiner-Neustedt, about twenty miles south of Vienna. It was to be another long trip, but thankfully not at low level. Heavy cloud cover obscured the land below as the bomber stream climbed to 22,000 feet. The men encountered the first resistance over a small town in Yugoslavia. In spite of a heavy German FLAK barrage, no American planes suffered damage. Once over the target, the airmen noticed hundreds of Messerschmidts lined up in rows as part of a military review. They quickly revised their plans, deciding to bypass the factory in favor of bombing the planes on the runway. After the attack, they began the long trip back to Africa. Over the Italian coast, they passed over a fighter base, where planes were sent to intercept them. A running gun battle ensued all the way to Sicily, but no bombers were shot down. When they had completed their mission, Titkemeyer and the other airmen learned that only one bomber had been lost; fortunately, its crew had been saved. Army Intelligence informed them that 400 Messserschmidts just off the assembly line had been destroyed. This mission gave Titkemeyer more satisfaction than the first. He hoped all future raids would turn out as well.

Three days later, on August 16, Titkemeyer started off early on his third mission; the target was a German fighter airfield at Foggia, Italy. A German JU88 followed the bomber stream at a safe distance, apparently radioing their altitude to the gunners on the ground. As the planes neared their target, antiaircraft fire began to burst all around them, shaking Titkemeyer's plane. Two B-24 bombers in their formation were hit and spun out of control, falling to the ground. No parachutes were visible. The remaining bombers, however, did not falter. They remained on course, released their loads, and made the turn for home. Then, German fighter planes appeared, apparently from the field below. The planes, swarming all around them, seemed to be everywhere. The guns of Titkemeyer's plane fired back at the attackers from the gun turrets. A bullet hit the right wing of his plane, puncturing the gas tank. Then the plane's glass dome shattered. But just as quickly as they had come, the German fighters were gone and the bombers limped home. Upon their return, the crew learned that seven of the twenty-four planes in their group had been shot down. Their plane was credited with downing one fighter. Overall, the raid destroyed forty-six German planes, thirty in the air and sixteen on the ground, and left the base in shambles. For Titkemeyer, however, there was no jubilation. His group's losses weighed heavily on him. He began to think that serving in the infantry would have been a better choice.

There was little down-time between missions, which was just as well from Tikemeyer's point of view. The base in the desert was excessively hot during the day, but at night after the sun went down, the temperatures dropped precipitously. It was little wonder that the men found it difficult to sleep. They bunked on hard cots in makeshift tents. The frayed sheets and thin blankets issued to the airmen did little to protect them from the bitter cold. Besides the extreme temperatures, they also had to deal with the wind-blown sand and the constant drone of planes. Further, it was impossible for them to keep clean with the sand whipping through the tents and penetrating their clothes. The rats, mice, scorpions, and centipedes, which seemed to be everywhere, added to their discomfort.

It was almost a relief when Titkemeyer was told to prepare for another raid on Foggia; this time the target was the railroad yards. At the briefing, the airmen were told that four groups of B-17s with a P-38 escort would be going in forty-five minutes ahead of them. Knowing there would be fighter protection, Titkemeyer felt more confident. The bomber left on time and dropped its bombs at 23,000 feet into the fires started by the lead B-17s. Although there was heavy FLAK, it was inaccurate, and none of the bombers was hit. No enemy fighters were encountered. Titkemeyer thanked the fighter escorts pilots personally when he returned to base.

Shortly after the second Foggia raid, Titkemeyer and his crew received orders to report for temporary duty in England. The long trip from their base near Bengazi to Marrakech, a distance of almost 2,000 miles, and from there to England, took several days. This was due to stopovers caused by poor weather conditions that reduced visibility to zero. For the next three weeks, in September 1943, Titkemeyer and his crew hit air dromes in occupied France. Intense FLAK and enemy fighter planes confronted them on virtually every mission, but they emerged unscathed.

In late September, they were ordered back to Africa, but to a new base near Tunis, with better accommodations and food. Titkemeyer's first mission from this base had as its target the railroad yards at Pisa, Italy. Visibility was good and no opposition was encountered. This was one of the few "milk runs" in which Titkemeyer was fortunate enough to participate. However, on October 1, Titkemeyer was ordered back to Weiner-Neustadt to bomb the same aircraft factory he and his men had been ordered to hit less than two months before. He dreaded this mission because he knew what to expect. From the very beginning, the operation did not go according to plan. Titkemeyer's bomber fell behind in the formation and had to fly on full power for three hours to catch up. Suddenly, at the I.P., several bombers turned and cut off his plane. Then, as they began the target run, the planes in the rear of the formation, Titkemeyer's included, became jammed together. The pilots of the B-24s became disoriented and the planes narrowly missed crashing into each other. At the same time, a tremendous barrage of FLAK tore through the formation. Less than two minutes before they reached the target, a swarm of forty enemy fighters swooped down on them, machine guns blazing. The lead American plane was shot down; in the heat of battle, the bombers dropped their payload before reaching the target and began to scatter. The No. 3 engine on Titkemeyer's plane was shot up and quit. The pilot had to feather an engine and, unable to keep up with the formation, the plane fell further and further behind. Titkemeyer watched in horror as several planes in his formation went down in flames. Meanwhile, the gunners in his plane fired at the attackers, with Titkemeyer firing his machine gun as well. He later recalled: "I was shooting at fighters all around us. Pretty soon one lined up and came right at me. I started pouring the lead at him but he kept right on coming. He got larger and larger. I got frantic and held down the trigger in a great long burst at him. Tracers were glancing off his engine. Soon smoke poured out of his engine and he dived into the clouds and was gone. I held down the trigger too long and completely burned the rifling out of the machine gun. I probably got the fighter." The other gunners aboard got four kills that day. After they had limped back to safety at Tunis, Titkemeyer said he had never been "so happy" in his life. But his joy was short-lived; he soon learned that only nine of the twenty-five bombers in his group had returned.

Less than a week later, Titkemeyer and his crew were ordered back to England to rejoin the 66th Squadron of the 44th Bomb Group to which they had been assigned previously. By then they knew the long hard route well. After an exhausting trip, they learned they had been named squadron leaders due to their experience and the losses bomber command had undergone. When he wasn't on missions, Titkemeyer flew on practice runs, teaching new recruits formation flying and other essentials of survival.

Titkemeyer's next missions were in the most northern part of Europe. After Africa, Titkemeyer was not used to the bitter cold weather, especially at high altitudes where the temperatures were usually fifty degrees below zero. On several runs, in spite of the heavy winter clothing, he and his crew got frostbite. Often the targets were obscured by overcast skies that prevented visual sightings. Moreover, the targets, especially those in Germany, were more heavily defended than those in southern Europe. However, an escort of numerous fighters gave the bombers greater firepower to deal with this threat. Because Titkemeyer was now the lead navigator of the entire bomber stream, he spent extra time studying his charts and maps and plotting the course and time to target. The day of the mission, he arose around 3:30 A.M. to review all the details yet again. Too many lives depended on him.

From November 1943, to March 1944, Titkemeyer and his crew hit military installations throughout Germany. Targets included Wilhelmshaven, Munster, Bremen, the Ruhr Valley, Solingen, Frankfort, Kiel, Hanover, Berlin, and classified missions to the French coast. In every instance, heavy FLAK was thrown up to stop them, but while they had numerous scares, they emerged with few scars. The days were endless, unrelenting, and terrifying.

The closest they came to ditching was the Bremen raid; their plane was almost crippled by heavy FLAK bursts which shredded their plane with metal slugs, severed the hydraulic lines so the landing gear had to be pumped down by hand, and tore holes in the bomb bay, in the No. 2 and No. 3 engines and the No. 2 propeller. German gunners also sprayed one wing with .30-caliber machine gun bullets. It was a close call, one that Titkemeyer said "aged him ten years."

On two occasions, Titkemeyer and his crew were given special briefings on important targets on the French coast. Allied intelligence had identified a series of secret German rocket installations that were under construction and hidden from view in barnyards and forests. Once completed, they would be capable of launching hundreds of shells into London daily. These installations had to be taken out, and quickly. Flying as lead navigator, Titkemeyer flew a course at 12,000 feet. Although the targets were obscured by FLAK bursts in the vicinity, the plane held course, straight and level, and the bombs were released over the target. Intelligence later confirmed the destruction of the sites. The second attack, with Titkemeyer again as lead navigator, was also successful.

Finally, on March 12, Titkemeyer was briefed for his final mission to take out a gun emplacement on the French coast. He later recounted his feelings: "When I was told I had one more mission to go, I was amazed. Until this time I never felt fear as I always thought that it was just a question of time before our crew would be killed. When I found out I had only one to go, I suddenly felt fear." Fortunately, this proved to be one of his easier runs. With Titkemeyer flying deputy lead, the bombing run was on target with no opposition. All the way home, Titkemeyer was more relieved than he had been in months. He was the first navigator in the 66th Squadron to complete all his missions. Several days later he was promoted to captain and sent to the 453rd Bomb Group. He spent the remaining nine months of active duty there as a tactical advisor with no additional flying. Ironically, his plane, "Myrtle, the Fertile Turtle" was shot down on its very next mission. Of the six bombers in which he had flown during combat, five were shot down and the sixth one was shot up, after which it was cannibalized for parts.

Charles Titkemeyer earned two Distinguished Flying Crosses, five Air Medals, and six Battle Stars. He remained in the Air Force Reserve for many years as a lieutenant colonel. After leaving the service, he earned a veterinary degree from Ohio State University and then went on to pursue his M.S. and D.V.M in Veterinary Medicine at Michigan State University. Following graduation, he spent the next twenty years at Michigan State, where he served on the teaching staff in the College of Veterinary Medicine. When Louisiana State University opened its new College of Veterinary Medicine in 1961, it selected Charles Titkemeyer to head up the school. Married for fifty-five years to his beloved wife Agnes, he passed away on July 17, 2002.

Maurice Alston

With tensions rising in Europe, Maurice Alston realized war was on the horizon. He graduated from high school in Mena, Arkansas, in 1938. He was only fifteen, but excelled academically, serving as salutatorian of his class. He then enlisted in the Arkansas National Guard. When Alston returned from maneuvers in Minnesota, authorities discovered that he was not yet eighteen, and he was discharged. But after he finally turned eighteen in November 1940, he rejoined the guard. His unit was called to active duty and he went through basic training at Fort Sill, Oklahoma. When he had completed training, he was dispatched to Louisiana in the summer of 1941 for several weeks of maneuvers. In early 1942, his field artillery regiment was shifted to Camp Bowie in Brownwood, Texas. While there, he applied for officers' candidate school at Fort Sill; he was accepted, graduating as a second lieutenant in March 1943. He was assigned to the 734th Field Artillery Battalion at Camp Maxey, near Paris, Texas. Several months later, the unit was ordered to Camp Gruber near Muskogee, Oklahoma. In the summer of 1943, Alston was sent to an Army Air Corps facility at Hatbox Field in Muskogee. However, after much deliberation, he decided he wanted to fly and applied for flight training. He did not hear anything and was sent back to Louisiana for additional maneuvers. When he returned from Louisiana in February 1944, he was thrilled to find his orders waiting for him. He was instructed to report to the San Antonio Aviation Cadet Center for classification and pre-flight training. He scored highest on the navigation section of the screening test, but elected to go into pilot training. He was sent to primary flying school at Sikeston, Missouri; however, he "washed out" in basic training even though he flew solo in primary school. Based on his aptitude score at the classification center, he was then sent to Selman Field in August 1944. After completing the course at Selman in January 1945, he reported to Kearns

Field near Salt Lake City, Utah, for overseas processing. He was next shipped to Travis Field, California. From there he embarked on the forty-four-hour trip to Biak, New Guinea, with fueling stops in Hawaii, Tarawa, and Los Negros. When his B-24 put down in Biak, he was greeted by the sight of the flag at half-mast; it was then he learned that the president had died.

Following an intensive two week LORAN course, Alston was assigned to the 70th troop Carrier Squadron, 433rd Troop Carrier Group at Leyte in the Philippines. The aircraft of his unit was a C-46, commonly called the "Curtis Commando." It was the largest twin-engine aircraft of its time, capable of carrying 50,000 pounds of cargo with a range of 1,200 miles. Its mission was to fly supplies and equipment wherever they were needed. As the island-hopping campaign of General McArthur continued to move northward towards Japan, the Carrier Squadron followed. The base of operations shifted continually, first to Clark Field on Luzon Island near Manila, and then to Okinawa.

On the night of July 28, 1945, Alston and his copilot were sleeping on the floor of the cargo plane when, about 2:30 in the morning, they were awakened by a tremendous crash. A kamikaze with a 500-pound bomb strapped under the fuselage came in undetected at a low level and slammed into the row of C-46s parked on the airfield. The right aileron of Alston's plane was hit, but the bomb stuck up tail first in the ground and did not explode. The pilot and crew chief were sleeping on cots under the aircraft tail and were awakened as the kamikaze plane came in. The men began running barefoot across the coral, receiving numerous cuts on their feet. Shocked and dazed, the men were thankful for what appeared to be divine intervention. Shortly thereafter, in early August, Alston and his crew were transferred back to Iwo Jima. There they took on a mission to carry fifty-five-gallon drums of gasoline to the occupation troops in Japan and bring out prisoners-of-war on the return trip. In mid-September, while they were in Japan, a fierce Pacific typhoon struck the Japanese mainland and wrecked Alston's C-46, together with eleven other C-46s and C-47s.

In October 1945, the 70th Squadron was moved to the Japanese mainland. Its mission continued uninterrupted: to haul cargo and personnel as needed. On one such trip, Alston's plane put down for refueling at a Marine airfield, where the Hollywood film star Tyrone Power was a marine pilot. He helped the crew refuel and autographed their $1 bills before they departed. By December, Alston had accumulated enough flying points to rotate home. He boarded the troop carrier *Leonard Wood* and arrived in San Francisco shortly before Christmas, 1945, almost five years from the time he had been called up.

Alston returned home in 1946 and applied for tuition grants under the G.I. Bill. He enrolled at the University of Arkansas, where he completed his first year. He then transferred to the University of Georgia in a pre-veterinary program, maintaining a straight "A" average during his one-and-a-half years of study. He was accepted into the College of Veterinary Medicine at Georgia, but toward the end of his first year. There, he was recalled to active duty in March 1951, during the Korean War. He remained in the service until his retirement in 1970.

Maurice Alston is the father of six children: five sons and one daughter. His wife, Zelma Boston, was his math teacher during his senior year of high school. After he graduated, the two had corresponded for a time until his mother put a stop to it. While Alston was on maneuvers in Louisiana in 1943, he was able to get a pass and go to Montgomery, Louisiana, a small town nearby, where the Boston family home was located. When Alston arrived, and asked about Zelma, Mrs. Boston misunderstood and thought he was looking for Zelma's younger sister, Dorothy. When Alston clarified the situation, Mrs. Boston replied that Zelma was living in San Antonio, where her husband was stationed. Two months later, when Maurice arrived in San Antonio for pre-flight training, he visited her. For years after that they remained in touch by telephone. After the war, Zelma and her family moved to Shreveport and Alston continued to phone her to say hello whenever he was in the area. They lost contact in 1968, after her listing was dropped from the Shreveport telephone directory. In 1975, when Alston flew into Barksdale Field in Shreveport, he went to the office of the *Shreveport Times* on a hunch. Zelma had previously been employed there and he wanted to see if they had a contact number for her. To his surprise, the receptionist called the circulation office, where Zelma worked and told her that a "gentleman" was there to see her. Zelma asked the receptionist to get the name of the visitor. Alston told the receptionist to tell Zelma that he was "no gentleman." Upon hearing this rejoinder from the receptionist, Zelma replied, "send him on in. I know who he is." By then, both he and Zelma had been divorced from their spouses. They began to date and were married in August 1976. Zelma passed away in December 2005, after twenty-nine years of marriage.

George Rorke

George Rorke was still in high school in New York City when he heard that Pearl Harbor had been attacked. During his senior year in high school, he took the ASTP, the Army Specialized Training Program Test, and the V-12, the Navy equivalent. After graduating from high school in June 1943, he registered for the draft. However, as he had not heard from the military, he found work in the Bendix Aviation Corporation in Brooklyn as a stockman. After working three months at Bendix with still no notice from the draft board, he enlisted in the Army Air Corps. He was processed and passed his physical; he was then ordered to board a train for an unknown destination. The train was headed for Miami Beach, where Rorke went through a leisurely basic training program. When he had completed the course, he was put aboard a train bound for Syracuse, New York. There, he was quartered in a fraternity house on the campus of Syracuse University and enrolled in English, history, and some military science classes. He also pulled guard duty and was provided with ten hours of flight training. After four months of coursework, Rorke reported to Maxwell Field, Alabama, which was serving as a transition-training base for B-24 crews. Here life was much more rigid and coursework and physical exercise were much more challenging. At the conclusion of pre-flight training, the cadets were given a series of aptitude tests to determine their future classification. Rorke scored well in the navigation section; he was sent to Selman Field, even though he had always thought he "would have made a good pilot." Instruction at Selman was very intense. Strict military protocol remained the order of the day, but Rorke adjusted to the demands. The time passed quickly and shortly before Christmas 1944, Rorke received his silver navigation wings and second lieutenant bars.

Rorke left Selman on a ten-day leave in New York City before reporting to a replacement depot at Greensboro, North Carolina. He was awaiting orders for overseas deployment when he received a telephone call from his mother. She notified him of his brother Thomas' death in an airplane crash in the South Pacific. Overcome with grief, he wanted more than ever to be part of the fighting so he could respond to the enemy in kind. Finally, after two weeks, his orders came to report to Taunton, Massachusetts, for transportation to England. From England, he was flown to an airfield in Beauvais, France, where he joined the 9th Air Force, 322 Bomb Group, 449th Bomb Squadron. The field was primitive. The airmen slept four to a tent, ate unappetizing "C" rations in a common mess, and sloshed their way from one tent to another in the mud.

Rorke was assigned to a B-26 medium bomber as a "toggleer." Navigators and bombardiers were assigned only to the lead planes in the formation. Consequently for all other planes, the navigators, acting as bombardiers, toggled the bombs when they saw the lead ship drop theirs. Only in the Army, Rorke thought, would men be trained as navigators with no training as bombardiers, and yet be used in that capacity.

Rorke went into action immediately, flying missions over Germany. Because he was in a medium bomber, the plane flew at 12,000 feet. Since it was late in the war, the Luftwaffe had been broken and Axis air losses were insurmountable. Therefore, little fighter opposition was encountered. Even anti-aircraft fire was weak or non-existent. On only one mission was enemy fighter opposition even sighted and that was most unusual. Rorke and his crew sighted their first combat jet fighter, a ME 262, but the squadron's fighter escort scared it off before it could engage. After the first few missions, Rorke and his bomb group were moved nearer the front, to St. Trond, Belgium. They occupied a German-built airfield, which had facilities, including a swimming pool, far superior to those found at Beauvais.

When Germany surrendered in May 1945, Rorke had not accumulated enough points, length of overseas service, or other considerations, to be discharged. He was transferred to the 305th Bomb Group in Chelveston, England; there, he was assigned to a B-17 group charged with photographing the European continent in detail. The planes were equipped with three cameras in what had previously been the bomb bay. Each camera was positioned in such a way that, when the film was developed, there was a panoramic view. All flights were to take place only in optimal weather conditions. Rorke's task was to chart the course for each day's flight. The group was soon transferred to a more central field near Landsberg, Germany, where it continued its mapping assignment for one year. In June 1946, Rorke received orders to return to the states for discharge. He sailed aboard the *Frederick Victory* out of Le Havre, France, for New York City, home, and final separation from the service.

Rorke married his long-time high school sweetheart, Catherine, in May 1947. During their fifty-one-year marriage, they had eight children and twelve grandchildren. Catherine passed away in 1998. Rorke tries to attend the Selman Field reunions to pay respects to his brother Thomas, whose name is inscribed on the Memorial Wall at the Selman Field Airport.

Theodore Homdrom

Theodore Homdrom, the fourth of five children, was born in western North Dakota in 1918. His grandparents on both sides had been born in Norway, as was his father. His father died when he was only three years old. His mother then moved the family to a small farm in northern Minnesota. As a child, he walked three miles to a one-room, one-teacher school. After graduation from high school, he attended Concordia College in Moorhead, Minnesota. Family finances were always inadequate, and Homdrom was forced to find work over the summers on farms or at other odd jobs to pay for his studies. Indeed, one year he was forced to drop out of college to earn enough money to stay in school. He graduated in 1941 with a major in history. He then accepted a position teaching Social Studies and coaching basketball at Comertown High School, Montana. His academic career ended abruptly with the Japanese attack on Pearl Harbor. In February 1942, he received his draft notice; he was inducted into the Army as a private in March of that year. His aptitude test scores indicated mechanical ability. He was, therefore, ordered to report to Camp Cook, California, where the 5th Armored Division was being organized and trained. He was pleased, as he had no desire to be in the infantry.

During basic training, Homdrom drove light and medium tanks. He enjoyed this assignment, having become proficient at double clutching while hauling grain during his previous summer jobs. On his twenty-fourth birthday, he was chosen to be a tank commander and promoted to PFC. He and his tank crew was then sent to the California desert for further training in circumstances that were supposed to simulate combat conditions in North Africa. But even maneuvers could be dangerous. One night, while he was asleep on the sand by his tank, Homdrom felt something move under his arm. He had been told that there were scorpions and rattlesnakes in the underbrush, and knew immediately that a snake must have been attracted by the warmth of his body. At first he froze, but then he jumped up before the snake could coil and strike. Running to the tank, he grabbed an axe; he then hurried back to where he had been sleeping and killed a rattlesnake, which was still coiled up on his blanket.

While on maneuvers, he was promoted, first to corporal and then to sergeant. One Sunday, while passing by regimental headquarters, he noticed a poster urging men to sign up for the Air Corps. The idea intrigued him. At least, he thought, in the Air Corps he wouldn't have to breathe in all the sand, experience the oppressive heat, or deal with poisonous snakes. He picked up an application, filled it out, and handed it in. Three months later, in November 1942, he was ordered to report to March Field, California, for the physical and mental examinations required by the Air Corps. A short time after returning to his desert base, he was recalled for an interview before the Aviation Cadet Board. With the Board's approval, he was ordered to board a train for the classification center in Nashville. After nine months of armored warfare training, Homdrom now prepared himself for an entirely new set of circumstances. Although he was not certain what to expect, he was ready for the challenge.

During processing at Nashville, Homdrom scored well on tests both for navigators and pilots. At first he was classified for pilot training, even though he had listed navigation as his preference due to the numerous math courses he had taken in college. However, because so few cadets had passed the navigation tests, there was a shortage of cadets for navigation school. Therefore, Homdrom was contacted and asked if he would consent to placement in navigation training. He agreed, and was soon on a train bound for Monroe. Although he had never been in a plane before, he knew he had made the right decision. Homdrom found both pre-flight and advanced navigation training at Selman, although challenging, to be exciting learning experiences. The ex-teacher was once again a student, interested in all facets of instruction. Graduation in August 1943, brought changes to his life. He was now an officer, no longer an enlisted man. His new status brought renewed responsibility, which he relished. After a fifteen-day leave at home in Minnesota, Homdrom reported to Moses Lake, Washington, to meet his B-17 crew. From there the crew was sent by train to Kearney Air Field, Nebraska, to begin flight training. Upon completion, the crew was then sent by train cross-country to a huge staging area near New York City. From there they were packed aboard the *Queen Mary* with 20,000 other servicemen for the transatlantic run to England. No escort vessels were provided for protection against German U-boats; to reduce the possibility of attack, the huge ship maintained the swift speed of forty knots, zigzagging every ten minutes to prevent German submarines from getting a clear fix on her position. Even though the rough seas and constant maneuvering made many of the men seasick, the ship broke all wartime records, making the crossing in less than five days.

Homdrom and his crew were processed and dispatched by train to Ridgewell Field to join the 381st Bomb Group, 8th Air Force. The veterans made them feel welcome, but the newcomers were disheartened to learn that they were needed to replace the many airmen who had been lost. The intensive orientation that followed included lectures on base living accommodations, flying conditions, as well as weather patterns in the British Isles. The airmen were also informed about the British air-sea rescue system in case a plane went down and the navigational aids they would need. One of the most interesting new advances was the top-secret British GEE Box, or ground electronic equipment; this was a newly developed navigational system that enabled the navigator to get a quick and accurate fix on his position. The new device was to be installed in all planes and under no circumstances was it to be allowed to fall into German hands. If the plane were to go down over enemy-held territory, the navigator was to push a button on the device, which would blow it up.

Early on Christmas Eve 1943, Homdrom was called into a briefing for his first mission. He was to be part of a huge assault force of 277 bombers dispatched over eight separate rocket sites in use or under construction along the English Channel. Homdrom's formation was ordered to attack and destroy V-1 launching sites at Calais. It was to be escorted by U.S. Thunderbolts and British Spitfires. Homdrom prepared his navigation flight charts and then put on his electrically heated flight suit. Although they would be flying at only 12,000 feet, the planes were not heated and the men could expect temperatures as low as –40 degrees Celsius. As his plane speeded down the runway, Homdrom was glad to finally be in combat after months of preparation; at the same time, he thought wistfully of the Christmas Eves he had once enjoyed. He breathed deeply and tried to ready himself for his trial by fire. He then prayed quietly "Lord, I am in your hands;" he would repeat this mantra during each and every combat mission in which he participated. Since the plane was not pressurized, he donned his oxygen mask at 12,000 feet. He then busied himself with recording flight observations in his log, plotting the course, getting fixes to make certain the plane was on the correct course, and reading the drifts to ascertain the wind speed. Homdrom and the crew encountered FLAK as they neared the target, but it was intermittent and inaccurate. However, Homdrom found it difficult to pinpoint the rocket placement due to heavy camouflage neetting. Finally, on the third pass, the plane dropped its bomb load and returned to base. Upon landing, Homdrom, thanked the Lord "for seeing me through this one." He was to repeat this prayer following every mission thereafter.

Homdrom's second mission came two weeks later, on January 7. Homdrom was awakened at 4:00 A.M. The target, he was told, was to be the I.G. Farben Chemical Works at Ludwigshaven, Germany. Although numerous enemy fighter planes were sighted, the American and British escorts engaged them in dogfights and prevented them from hitting the bomber formation. FLAK over the target did bring down one bomber, but the rest of the formation returned safely to base.

After Gen. Jimmy Doolittle assumed command in early January 1944, the tactics of the 8th Air Force underwent a change. General Doolittle targeted aircraft production plants and airfields in Germany as his primary objective. The initial attacks were to be followed up by further assaults that increased in intensity until the Luftwaffe was decimated. On January 11, Homdrom was ordered to be part of a massive strike force of 800 bombers, accompanied by several hundred fighter escorts. Their target was the A.G.O. aircraft fighter assembly facility at Oschersleben, ninety miles southwest of Berlin. This was to be the deepest penetration ever made by allied aircraft into the German heartland.

After the bombers had crossed the Channel, a weather front approached England. Bomber Command ordered the bombers and their escorts to return to base; however, 139 bombers did not receive the order and continued to the target with only a few fighter escorts. The results were catastrophic. Swarms of enemy FW 190s pounced on the group. Machine gun chatter drowned out the roar of the bombers' engines as the air fight raged on. A running gun battle ensued; it lasted over one-and-a-half hours, continuing on the approach to target, over the target, and for an hour afterwards. Forty-two bombers, including eight from Homdrom's group, were shot down.

For the survivors, to lose so many men was devastating. Army Intelligence officially credited the combat wing with shooting down sixty-six enemy fighter planes and damaging another 145. Homdrom's group alone was credited with downing twenty-eight enemy planes; but this did not ease the sense of loss felt by the airmen. After the mission, Homdrom's group was told that this had been the heaviest enemy fighter plane opposition had been the heaviest in the war up until that time, and it was awarded the Presidential Distinguished Unit Citation.

On February 3, Homdrom was in the air again; this time the destination was the key port of Wilhelmshaven, Germany. While over the target, one of the plane's engines took a direct hit from FLAK. As the plane began to lose speed and altitude, it dropped further and further behind the formation. The crew "sweated out" their return, knowing German fighters often attacked a lone bomber. Fortunately, the crew returned to base safely, much to the joy of the other crews and ground personnel who anxiously awaited them, fearing that they had been lost.

On the following day, February 4, though still not fully recovered from the strain of the previous day, Homdrom was awakened early and told to get ready to fly. His crew was to head to Frankfurt, the site of railroad marshalling yards and manufacturing plants. During the mission, the FLAK was again intense, but no bombers suffered any direct hits. Fighter escorts prevented German warplanes from disrupting the raid and attacking the bombers. All crews returned home safely. Homdrom was ready for a good night's sleep.

However, his respite was brief. On February 6, he was part of an attack on airfields in Nancy, France. Due to heavy undercast, the bombers could not see their primary target and bombed a secondary target instead. Enemy fighter opposition was intense, but the bombers maintained their formation and returned to base with no losses.

On February 11, Homdrom was again awakened early to be briefed on the day's raid. This was to be a return to Frankfurt and the railroad marshalling yards. While the flight to target was uneventful, Homdrom was concerned, as their pilot was maneuvering the plane unusually close to the next bomber. While this would give the group added firepower in the event of an enemy fighter attack, it used up precious fuel. Then, over the target, the bombers had to climb even higher to avoid the contrails of those bombers that had preceded them, again using extra fuel. On their trip back to base, the pilot realized there was not enough fuel left to reach England. He ordered the crew to pitch anything heavy out of the plane, including their guns and ammunition. They prepared to ditch the plane in the frigid waters of the North Sea and contacted Air Sea Rescue. As they approached the White Cliffs of Dover, they noticed a flat green meadow just beyond the cliffs. They decided to go for it. What they could not see from that distance was the barbed wire, electric lines, and an eight-foot wide ditch that crossed the meadow. The plane plowed into the field, slamming into the ditch. Everyone survived, although the ball-turret gunner suffered two cracked ribs. The next day Homdrom wrote his mother: "I'm beginning to wonder if I could settle down to a quiet life of teaching again."

Homdrom did not fly again for eleven days. He used the downtime to secure a pass and travel around England. Due to his interest in education, he visited Oxford and Cambridge universities, toured London, and tried to see as many historic sites as time would allow. He was impressed with the beauty of England and the kindness and courage of the British people. But even more important, the time away permitted him to unwind and for a brief instant to escape the constant tension.

When he was back on base, he joined in a chorus composed of any willing and available airmen which sang at Sunday morning church services. He had sung in the choir at Concordia College; participation made him feel comfortable, for it reminded him of his life before the war. He looked forward to participating in these services whenever he was not on duty.

During the third week of February 1944, Homdrom went on two missions. On February 20, his target was the aircraft factory at Leipzig; this was one of only three sites which were to be attacked by over 1,000 bombers. Intelligence reported good results at all three targets. On February 24, Homdrom's target was the strategic ball-bearing factories at Schweinfurt. Previous raids, the men were told, had resulted in substantial bomber losses, but this time new long-range P-51 Mustangs were to be deployed. This increased security blanket gave the airmen an added morale boost. Nevertheless, the crew knew that flights that deep into Germany would trigger fierce opposition. Yet, in spite of heavy FLAK and head-on strafing attacks by Luftwaffe fighters, the mission achieved good results. It wasn't until the crew touched down, emotionally drained and physically exhausted, that they realized how close they had come to disaster. They counted over 1,000 holes in their plane from FLAK and machine gun bullets, including a huge hole in the tail gunner's turret. When they asked the tail gunner what had happened, the full-blooded Hopi Indian from Arizona replied that just as he was reaching back for more ammunition, a German fighter's 20 mm shell exploded right where he had been. The crew had dodged yet another close call and chalked up another day of heavy combat.

The two raids on Leipzig and Schweinfurt were part of a massive effort by bomber command to cripple Germany's war-making capacity. The ongoing assaults on oil refineries, bridges, railroad marshalling yards, supply trains, submarine pens, and war industries were carried out relentlessly day and night. The Allies put up 5,100 bombers: the RAF 1,300, and the United States 3,800. In addition to German losses on the ground, 450 Luftwaffe planes had been shot down, numbers which Germany could not endure much longer.

Following the Schweinfurt raid, a wintry mix of ground fog and intermittent rainy weather grounded most of bomber command for a week. Then, on March 2, Homdrom and his crew were ordered out again, this time back to Frankfurt: their target was the Alfred Ternes Aircraft component factory. Moderate opposition did not deter the bombers, and the strike force achieved good results.

After this mission, Homdrom had little time to rest. Early the next morning, he was ordered on a mission to Oldenburg; however, due to poor visibility, the bombers had to fly up to 28,000 feet to get above the clouds. On the ascent, two of the bombers collided. It was impossible for the other bombers to get back into formation, so the bombers were redirected to a secondary and closer target, the port of Wilhelmshaven.

Homdrom kept hoping for some downtime to ease his constant stress, but bomber command was determined to destroy the Nazi war machine before the expected invasion. So on March 4, for the third day in a row, Homdrom was awakened early and told to report to a briefing concerning a raid on Berlin. Again, a weather front caused the bombers to abandon their primary target in favor of the secondary target of Dusseldorf, where many war-related factories were located. Intense FLAK over the city brought down one bomber, but the rest were able to return safely to base.

Homdrom had only one day of rest before he was called upon again to participate in another raid on Berlin. By this time he had become conditioned to early morning wake-up calls. He knew that any strike on the German capital would be met with fierce oppostion. The briefing officer informed the men that this was to be a large-scale assault of almost 700 bombers. All the men felt the apprehension. After the briefing, one crewman was overheard saying "the queues to the toilet are three times as long today."

As the Allied planes approached the target, thirty-six German single-engine ME 109s tore through the fighter escort and hit them hard. Swirling dogfights broke out everywhere. Aboard the plane, every man who had a gun, including Homdrom, opened fire on the Germans. Homdrom had a .50-caliber machine gun at each side window and raced back and forth firing. As he fired on one plane bearing down on them with guns blazing, it veered and he saw it crash into a nearby B-17. A huge fireball erupted as both planes fell out of the sky. Other ME 109s burst into the formation shooting down the B-17s on Homdrom's right and left. Machine gun bullets and rocket fire peppered the squadron, but the bombers held tight to their course as they flew through the growing carnage. Homdrom uttered a silent prayer: "Lord, if you see me through this, I'll do what you want me to do." Somehow the remaining planes dropped their bombs over the target and returned to base. But for almost five hours, over the target as well as on the return trip, there was a running gun-battle with German fighters. The raid cost bomber command sixty-nine bombers. Each side also lost 100 fighter planes as well. Homdrom's crew was credited with shooting down four enemy planes.

Two days later, on March 8, Homdrom was again awakened and briefed for yet another raid on Berlin, this time on the Erkner ball-bearing works. He felt shell-shocked and did not know if he could survive another close call. His only sustenance was prayer. As always, when his plane left the runway, he prayed, "Lord, I am in your hands." The weather was good, and the bombers from various bases linked up and assembled in formation over the English Channel. Numerous fighter plane escorts joined them and formed a protective shield around the formation. FLAK and enemy opposition were not as lethal as before, but resistance was still fierce. Indeed, the planes were under attack for almost five hours. Thirty-seven bombers went down that day.

The stress of accrued combat missions was taking its toll. Therefore, when Homdrom was offered the chance to get away following the raid, he gladly accepted. He decided to go to London with a friend. They tried to reserve a room at the Regents Street Hotel, in the heart of the theater district, but, as there were no vacancies, they settled on a room near the train station. While they were having lunch near Picadilly Circus, an air-raid alert sounded. Suddenly the area was jarred by a huge explosion. Homdrom learned later that a V-2 rocket had scored a direct hit on the hotel where he had unsuccessfully tried to make a reservation. Homdrom hoped he would get another chance to attack the rocket launching pads and pay the Germans back for the death and destruction they had caused.

As D-Day drew near, the pace of attacks became even more relentless. Between March 20 and March 28, 1944, Homdrom was ordered up on five missions. By this time, after only three months in the combat zone, due to numerous losses, he had become the second most experienced navigator left in the squadron. On March 20, the target was Mannheim, Germany, a key industrial complex. The bomber wings, fifty-four bombers strong, were propelled to the continent by a strong tail wind; at the same time, the contrails from the bomber wings in front of them resulted in a heavy cloud formation, forcing the wings in the rear to climb to 30,000 feet. As they dropped their bombs over the target, the men were horrified to see another wing formation flying right at them. Bombers struggled to avoid one another; Homdrom and his crew watched in shocked disbelief as two nearby planes crashed into each other. The order and discipline of the formation rapidly disintegrated.

Homdrom's plane found itself flying alone. The pilot asked Homdrom for a heading back to base and then proceeded as he had indicated. But, seeing a bomber formation following a pathfinder off in the distance, the pilot changed course to join the others. Homdrom, however, questioned the course the formation was taking. Pulling a small map of Europe out of his briefcase, he scoured it for significant landmarks. Peering down between the clouds, Homdrom made out a river flowing due west. He realized that could only be the Loire River. Homdrom became convinced they were far off course and heading to the Bay of Biscay off the western coast of France. He called to the pilot to change to a heading due north and the pilot duly altered course. Over a dozen planes in the formation veered off to follow them. After being in the air for over eleven-and-a-half hours, they landed safely in England. Homdrom felt this was his finest job of navigating. However, the entire episode became known as the worst navigational error in the annals of the European Theater of Operations. A few bombers managed to land in neutral Spain. The rest of the formation literally disappeared, never to be heard from again.

Four days later, on March 24, the orders were to go back to Schweinfurt for a second time. One bomber crashed shortly after take-off and, as the planes climbed through the clouds, two bombers collided. Three bombers and thirty men were gone before they had even encountered enemy action. The culprit that day was the weather, which proved as daunting a foe as the enemy. There was little talk among the crew as they crossed the Channel. Each man was deeply immersed in his own thoughts; the men prayed they could complete the mission without further casualties. Meanwhile, weather conditions worsened and the planes were redirected to the secondary target, Frankfurt, and the war industries located there. While bombing results were satisfactory, none of the crew could find any joy in the mission. The cost had simply been too high.

Two days later, on March 26, Homdrom received the welcome news that he was to participate in a mission to bomb the German V-rocket bomb-launching sites in Calais. To Homdrom, this would be retribution for the havoc he had witnessed at the Regents Hotel less than three weeks before. It also took his mind off the tragic losses incurred on his last missions. This raid was successful; although many planes were hit by FLAK, all planes returned safely to base.

With good weather on the following day, March 27, Bomber Command dispatched 290 bombers to attack airfields in occupied France. Homdrom was part of this assault force. His target was the airport at St. Jean d'Angély, France. The next day Homdrom was ordered to Reims, France, to participate in the bombing of its airfields. By now, the airmen realized that the rumored invasion was not far off and the priority, in order to be successful, was to deny the Germans air bases near the landing beaches. These raids helped to boost morale among the airmen. They knew they were doing their part, no matter how high the cost, to weaken the German war machine.

As March became April 1944, the strikes continued with little pause. On March 29, Homdrom was part of a 236-plane formation that attacked Brunswick, Germany, a highly industrialized city with numerous war industries. For his next mission, Homdrom was informed that due to his experience, as well as the high losses, he was being appointed squadron navigator. He was to fly with the lead crew and would be heading a wing formation. The wing was comprised of three groups. Each group consisted of eighteen bombers with 180 men for a total of fifty-four bombers and 540 men. It was a huge responsibility, but Homdrom knew he was capable and was confident he could succeed. The targets of his first mission as squadron navigator were the airbases in the Brussels area. The mission was successful.

On April 11, 1944, Homdrom was told at the early morning briefing that the day's target would be the jet-fighter factory at Cottbus, Germany. This was a highly classified priority target, which had to be destroyed to keep Germany from putting its new secret weapon into widespread production and possibly altering the balance of power in the air war over Europe. Again, as squadron navigator, Homdrom flew in the lead plane of the fifty-four bomber formation. Heavy cloud cover over the target made it difficult to identify the production factory. Therefore, the group flew over the target and made a 360-degree turn. As the plane approached the target a second time, the bombardier in Homdrom's plane yelled over the intercom that he could see the target through a patch in the clouds. The pilot told him to go ahead and bomb it. As the bombers flew past, Homdrom saw the entire area burst into flames and smoke. Aerial photography later confirmed the destruction of the jet plane factory. Homdrom's group was awarded the Presidential Citation for this action. The destruction of the factory at Cottbus helped ensure the Allies' European air supremacy.

On April 28, Homdrom was again awakened early; after breakfast, he was informed at the briefing that he would be leading a mission to blast the airfields at St. Averd, France. He prepared his flight plan and went to the airfield to await take off. There he was notified he would not be going on this mission. Instead, he would be flying to Berlin at another time. Several hours later, as was the custom, he waited on the flight line with the other airmen for the planes to return from France. It was then he was told that the lead bomber he would have been flying in had taken a direct hit over the target and his replacement navigator had been killed. Again, he said a prayer; while deeply saddened, he realized how fortunate he was to still be alive.

The mission to Berlin to once again bomb the Erkner ball-bearing plant took off on April 29. While preparing his flight plan, Homdrom, who kept a "flak map" which indicated those areas where he had run into heavy FLAK on previous missions, noted that they would be flying over a "hot spot." Thirty miles from the target, he advised the pilot to alter course to avoid the expected assault. The pilot followed his advice and flew out of range of the heavy anti-aircraft placements. Fighter escorts prevented German aircraft from hitting the fifty-four bomber formation, and the FLAK was moderate. The bombing results appeared good. More importantly, no bombers were lost. On his return to base, the colonel congratulated Homdrom for the diversion he had suggested. His initiative had been at least partly responsible for the success of the mission.

On May 12, Homdrom again led the wing in an attack on industrial sites in Leipzig. More long-range P-51 mustangs came on line to provide fighter protection and again, the results were satisfactory, with no bombers lost. When Homdrom had arrived in England a year earlier, he had been informed that twenty-five missions would be the maximum for the flight crews before they were rotated home; since that time the number had been raised to thirty. Now, after all he had been through, he seriously questioned whether he could survive five additional missions. He was now a seasoned veteran, but he knew his good luck could run out at any time.

On May 28, he was again sent back to Leipzig as lead navigator. This time the target was a refinery but haze and smoke obscured it and prevented the bombardier from getting an accurate fix. Therefore, the planes bombed the secondary target, an airfield. Fighters escorts prevented the loss of any planes. Following the raid, Homdrom was greeted with the good news that he had been granted a seven-day leave. He spent much of his time at a country estate administered by the Red Cross. The rest, beautiful English countryside, good food, and fellowship lessened the stress. But he still could not forget that he needed four more combat missions to complete his tour; this reality continued to weigh on him.

On June 2, he flew with the lead crew of a fifty-four-bomber squadron. Rumors were swirling of the impending invasion. All passes had been cancelled; the targets were all in France. Homdrom's wing was assigned airfields near Paris at Boulogne and Paliseau. Thanks to good visibility, the results of the bombing were excellent. At 10:00 P.M. on the night of June 5, the lead pilots, navigators, and bombardiers were called to group headquarters for a briefing. This was unprecedented on the night before a mission. Everyone sensed that the invasion was imminent. As expected, the men were told that the invasion was about to begin. They were to fly in a narrow six-mile wide corridor at 14,000 feet, the lowest altitude Homdrom had ever flown on a bombing mission. They were to hit coastal defenses. Their aim and timing had to be perfect, since Allied troops would be storming the beaches immediately after the bombs were dropped. Uncharacteristically, they were issued pistols in case their planes went down and they had to engage the enemy.

An air of excitement swept over the room. The day, so long anticipated, had finally arrived; the airmen were overjoyed to be part of this historic mission. There was little time to sleep. Homdrom went to work studying his charts and thoroughly rechecking his flight plans. There was no margin for error. After an early breakfast the next morning, he went back to his charts and maps for one final check. Once the planes were in the air over the English Channel, he peered down to try to catch a glimpse of the invading armada, but cloud cover made visibility nil. Then through a small opening in the clouds, he saw a huge fleet of all types of ships moving toward the French coast. He was astounded by the massive size. Before he knew it, he was over the strike zone; the bombers dropped their payload and then turned for home. It was the shortest mission he had ever been on, but he hoped the role he played helped the men below establish a beachhead.

On June 14, Homdrom was once again sent out as lead navigator to bomb airfields in the vicinity of Paris. Visual sightings after the raid indicated the airfields had been badly damaged. All aircraft returned safely. Homdrom had now completed twenty-nine of his thirty missions. Somehow and for some reason he had been spared up until this point. He always worried before a mission, but now his apprehension increased exponentially.

On June 21, he was briefed for his final mission, a raid on Berlin, or as the airmen referred to it, the "Big B." The target was Luftwaffe headquarters and other government buildings in the heart of the city. He was to fly wing lead one last time. Expecting that FLAK would be extremely heavy over the German capital, he carefully prepared his flight plan during the three hours prior to take off, as was his custom.

At first everything went according to plan. The bombers grouped themselves in formation and escort fighters from various bases began to rendezvous with them. Before all the escort fighters could arrive, however, they were attacked by over fifty German fighters. Suddenly smoke and flames erupted everywhere, as twenty-millimeter shells and machine gun fire exploded all around them. Several B-17s were blown out of the air. Compounding the danger was the lack of visibility caused by dense clouds and the contrails from previous groups. Nevertheless, the bombers did not deviate from their course and the formation stayed together. Forty miles from target, Homdrom notified the pilot that they had arrived at their initial point, and the pilot called on the bombardier to take over. The autopilot was then switched to the Norden bombsight. Homdrom rechecked his flak vest and put on his steel helmet. Then he stood up to look over the bombardier's shoulder and watch him zero in on the target. At that moment, FLAK burst all around the plane. A thick piece of metal tore through the plane's bottom and shot through the top of the Plexiglas nose, missing the bombardier by inches. Homdrom covered his left eye with his hands and an instant later a second chunk of FLAK smashed through the bottom of the window a foot from his head. The shattered glass flew everywhere, slicing the back of his hand, his neck, and damaging his right eye. Although his vision was blurred and his face bloodied, he managed to give the pilot a direct heading back to base. The pilot then tried to transfer from auto pilot back to the plane's regular controls. However, the control cables had been torn apart by the FLAK and the plane was forced to remain on autopilot. Homdrom's plane was not the only one to have been hit by enemy fire. En route home the crew learned that twenty-seven of the remaining thirty-seven planes in their group had been hit by heavy FLAK.

As they approached England, the pilot notified the crew that he could not guarantee a safe landing on autopilot. He told them over the intercom that they could either bail out or risk an emergency landing. The crew decided to stay with the ship. Homdrom held his breath as the pilot brought the bomber for a safe landing. Homdrom was rushed to the base hospital. There he was told that fortune had smiled on him yet again: only the white of his eye had been cut. Later, while recovering in the base hospital, Homdrom received his second Distinguished Flying Cross as well as the Oak Leaf Cluster and Purple Heart.

After his release from the hospital, Homdrom prepared to go home. During this time, while at the officer's mess, the actor Edward G. Robinson, who was touring the airbase, saw him sitting with friends. Since Homdrom's eye was still bandaged, Robinson walked over to the table to ask about his injury and the two men engaged in a brief conversation. This encounter boosted Homdrom's morale. He marveled at the fact that he had defied the odds and survived. Since he had joined the squadron, twenty-seven navigators were no longer on the roster to fly combat. Only four had completed their tour, and of those, three had received Purple Hearts. Homdrom knew he had needed a miracle to survive, and he felt he had been granted one. Just knowing he would be leaving the war zone with its ongoing stress and danger enabled him to sleep soundly for the first time in months. The next morning, he said his good-byes to the officers and men with whom he had served.

Upon his return to the States, Homdrom's first visited New York City and his parents in Minnesota. When his leave was over, he boarded a train for Miami and reassignment at the redistribution center. His two-week stay in Miami awaiting orders was like a vacation. The luxury hotel where he was booked catered to returning veterans and provided them with the rest and relaxation they needed to transition to combat-free duty.

Homdrom's first assignment was Selman Field in Monroe as an instructor. So he had come full circle. This time, without the pressure of studies, he got out into the community more, met people, and enjoyed an active social life. He was disappointed when he received a reassignment to the navigation school in San Marcos, Texas. He had looked forward to winding up the war in Monroe.

In December 1944, Homdrom arrived in San Marcos as an assistant flight commander assigned to fly with the students. After only three weeks, the base commander went on leave, and Homdrom was left in charge. In mid-March 1945, Homdrom and other veterans were ordered to the officers' command and training school in San Antonio. He had just finished the course when news was received of the president's death. Less than a month later, Germany surrendered. Because there were sufficient airmen in the Army Air Force to finish up the war in the Pacific, the Army's discharge system allowed personnel with sufficient points to apply for discharge. Homdrom knew he did not want to make a career of the military, and he did not want to chance surviving another tour of combat duty. Therefore, at the age of twenty-six, after three-and-a-quarter years in the military, he left the service.

Still not certain of what he wanted to do, Homdrom enrolled at the University of Minnesota in its graduate program in history under the G.I. Bill. There, he met Betty Stenberg. The two were married in June 1946. That fall Homdrom took a teaching job at Ellsworth, Minnesota, where he also coached basketball. After he had taught one year, Betty gave birth to their first child, a son. At the same time, Homdrom made a decision he had been mulling over for many months: he applied to a Lutheran seminary. While he was still in the seminary, his second child, a daughter, was born. When he had completed his training, he was asked if he would consider serving overseas. He accepted, and in November 1950, Homdrom and his family began thirty-five years of mission work in South Africa. Homdrom, therefore, kept his promise to God to do his will, should he survive. Since their return, the Homdroms have resided in St. Paul, Minnesota.

David Manheim

David Manheim was raised in Syracuse, New York. Even as a child, he was captivated by planes and flying. He would watch planes that passed overhead and spend hours making model airplanes. As a teenager, he persuaded a friend of the family who owned a plane to teach him the basics of flying. Manheim was enrolled in the Ohio State University School of Dentistry when the attack came on Pearl Harbor. He left school immediately, volunteering for the Army Air Corps. He was sent to Carlson Field in Arcadia, Florida, for basic and preflight training. During instruction, he suffered a minor crash on the runway. No one was injured, but the crash on his record prevented him from flying missions overseas. Disappointed, he reluctantly accepted assignment to Selman Field as an instructor. His skill and flying experience were to serve him well in this capacity.

After his arrival at Selman, he was befriended by a local couple, Rose and Jake Pickens. They loaned him 25 cents and lightheartedly told him they expected it back when he left Selman.

Later, while scanning the list of incoming cadets, he was thrilled and surprised to see the name of his younger brother Charles. Manheim was able to secure a room assignment with Charles while they were both at Selman.

In 1945, Manheim was transferred to Hunter Field in Savannah, Georgia. While he was on a training mission, his plane crashed into the ocean, killing the other two crew members. Somehow Manheim managed to extricate himself from the plane and put on his life jacket. He floated for five hours in the icy water until a young girl spotted him and hauled him aboard her boat. Manheim was severely injured, suffering forty-eight compound fractures. He was hospitalized for two years.

When news of the crash reached Selman Field, the Pickens drove all the way from Monroe to Savannah to visit him in the hospital. They told him that, due to the circumstances, he could keep his quarter.

Steven E. Schanes

Steven Schanes was raised in Orange Park, New Jersey. After graduation from high school, he attended Montclair State University. Two-and-a-half years later, he left to join the Army Air Corps. His transition from civilian to military life was facilitated by the ease of his early assignments. In February 1943, he went to the Pennsylvania Railroad Station in Newark boarded a train to Atlantic City. There, he was bussed to a large hotel, where he had a brief orientation and where he was to stay. He was given military clothes and, for the next four weeks, he had rifle practice and did daily calisthenics. It was a fairly stress-free basic training with surprisingly comfortable accommodations.

His good fortune continued. From Atlantic City, Schanes was sent by train to Penn State University, where he was lodged in a vacant fraternity house. For the next six weeks, he was enrolled in physics, mathematics, and English communication courses, but there was ample free time to enjoy campus activities and socialize. Indeed, Schanes' concerns about military life were fast disappearing in this relaxed atmosphere. However, life was about to change.

Schanes was next sent to the Nashville classification center, where he remained for six weeks doing KP duty and marching in daily drills. On the basis of his test scores, he was shipped off to Maxwell Field in Montgomery, Alabama, to begin pre-flight pilot training. Although his test scores qualified him for pilot training, he knew this was not what he wanted. He arrived in Montgomery in July, during the height of a steamy hot summer for which he was totally unprepared. The humidity drained him during the rigorous physical exercises, long marches, and parade drills. The classes and study time were rigidly structured and a strict military atmosphere prevailed. Schanes was now having serious second thoughts about the future the Army had in store for him.

When he had completed training at Maxwell, Schanes was sent by train to Union City, Tennessee, to begin primary flight training. The camp was administered by civilian instructors intent on teaching the students to be good pilots. The instructors were not interested in military discipline, so between training flights, there was ample free time. Solo training flights took place in sleek, new Fairchild PT-19 single engine planes. However, the more Schanes flew, the more certain he became that he did not want to be a pilot. He had never liked heights, got motion sickness at higher elevations, had difficulty in determining distances from the air, and could not differentiate left from right. When the end finally came and he was informed that he had washed out, he received the news with a sense of relief. As he later recalled, it was one of the happiest moments of his life.

Schanes was then sent to gunnery school at Panama City, Florida, in November 1943. Even though he did not feel totally confident, he was able to complete the course and receive his aerial gunner's wings. From gunnery school, Schanes was sent to a twelve-week course in navigation at Selman Field. While he could master the classroom instruction, Schanes found that his old problems resurfaced during the time spent in the air practicing navigational skills. However, although he never overcame his "directional dyslexia," he was able to master his constant motion sickness thanks to the seasick pills his mother sent him. At higher elevations, once he donned his oxygen mask, his fear of heights diminished. In March 1944, he received his gold bars and navigator's wings. He was assigned to combat crew training in Tonopah, Nevada.

Following a long train ride, Schanes arrived in Tonopah; there he underwent combat flight training from April to June 1944. This was the first time he had seen the B-24 Liberator, the plane in which he would be training. He met his fellow crew members who proved to be much younger than he had expected; the oldest, the pilot, was twenty-eight, and the youngest, the gunner, was only eighteen years old. However, over the next three months of flight training they sharpened their individual skills and the chemistry among them jelled. As their missions continued, each crewmember underwent flight checks by trained observers. Various simulated bombing runs were made, including one over Long Beach, California, in which they were "attacked" by Navy corsairs. As the training intensified, it became evident to the crew that they were being prepared for war on the Pacific front. They were given study assignments on identifying Japanese planes and naval vessels; in addition, they were told to buy light clothes for tropical weather and to be prepared for long flights using celestial navigation. When they had completed their training, the crew was ordered to Hamilton Field, California, to pick up a new B-24 and await their orders.

Schanes and the crew conducted thorough flight checks and flew several trial flights in their new plane. After several days of testing, they took off for what was still an unknown destination. Once airborne, they opened their flying orders and were stunned to see that their destination was Great Britain. The trip was not an easy one, but it was broken up with stopovers in Lincoln, Nebraska, and Goose Bay, Labrador. On the next leg of their journey to Reykjavik, Iceland, they were forced by rough weather to divert to Greenland. They made a difficult landing in a narrow valley and, after a brief respite, prepared to take off yet again. Fortunately, the flight to England the following day was uneventful. They were then shipped out to Northern Ireland for additional training. Here they studied German fighter planes until they were able to identify them. They also were given accelerated instruction on European geography and provided access to the latest advanced technical equipment, jamming techniques, and the new British "G" system of navigation. By September 1944, the crew members had completed their intensive study; they were assigned to the 93rd Bomb Group, 8th Air Force, in Harwick, England. They were soon made aware of the fact that they were a replacement crew due to the catastrophic losses the 8th Air Force had suffered; over twenty-five bombers had been lost. The likelihood of the crew completing its required thirty missions seemed slim. However, the men knew their preparations were thorough; while nervous, they were nevertheless confident.

Schanes and his crew flew deep into the heart of Germany in "Nikki," named for the pilot's girlfriend. As they bombed airfields and other war-related targets, their B-24 was shot up numerous times. They were attacked by enemy fighter planes on several occasions and watched many friends perish in the air war over Europe. On every bombing run, conditions in the plane remained uncomfortable. Since the bombers were not heated, temperatures of –55 to 60 degrees Celsius were commonplace at high altitudes. Schanes wore long underwear, a bulky flight jacket and pants, and heavy fur-lined gloves with thin silk gloves underneath to better operate his instruments when he had remove his heavier gloves. Nevertheless, he was always cold. When he donned his oxygen mask and cumbersome backpack parachute, it was even more difficult to maneuver in the narrow confines of the plane. After he was issued a new electrically-heated flight suit to replace his bulkier clothes, he had more freedom of movement which gave him a huge sense of relief. Following every mission, the crew was greeted by Red Cross girls who patiently awaited their return. They offered the crew hot coffee and conversation. The crew always looked forward to seeing them.

Then, in December 1944, just three weeks before he turned twenty-one, Schanes endured a particularly grueling bombing mission; afterwards he was promoted to first lieutenant. Later, he received the Oak Leaf Cluster to add to the Air Medal he had been awarded for an earlier mission; he felt proud of his accomplishments.

It was his thirtieth and last mission in February 1945, that Schanes always felt helped win the war: this was the raid on Remagen. Schanes served as the lead navigator on a thirty-six bomber strike flying in at low level, roughly 10,000 feet; their mission was to destroy the bridge over the Rhine and close the escape route to fleeing German troops. Schanes saw the bridge clearly on the approach; indeed, it presented a perfect target. The bombardier scored a direct hit on the bridge, completely destroying it. Later the crew discovered from aerial photos that they had destroyed the autobahn bridge, and left the railroad bridge, which was two hundred yards away, standing. It was the railroad bridge that advancing American troops were to use shortly thereafter to establish a bridgehead over the Rhine.

Roy Gates

Roy Gates, from West Monroe, joined the Army in 1942, when he was twenty-six. He had previously served two hitches in the Louisiana National Guard. He was sworn in at the downtown post office along with a number of other local enlistees. Later the group's induction photo appeared on posters around the community to encourage more locals to enlist.

Gates was sent to Fort Humbug in Shreveport for his induction and physical examination. He was next sent to Camp Beauregard for processing, immunizations, and further tests; here he was also issued military clothing. He was then sent back to Monroe by train with seven other Monroe-area boys and told to report to Selman Field. He thought this was rather ironic; he felt he had come full circle because, as a teenager, he had hunted doves and rabbits on what was to become Selman Field. After his arrival, Gates called the base but was told there was no place for him and the others to stay. He was told to go home and report back the next day. The following day they were put to work assembling bunks in the barracks.

Before he had enlisted in the service, Gates had tested instruments for United Gas. Since the flight line needed trained instrument specialists to check the instrument panels on all the training planes at Selman, Gates was assigned to the flight line, 96th Squadron. His job was critical because the cadets had to rely on their navigational instruments. The instrument panels were mounted on rubber shocks and the vibration of the planes often caused the wiring to come loose. Faulty instruments could ground the planes and delay training or, at worst, cause accidents. Gates was charged with checking for loose wires and fixing electrical shortages. He remained on the flight line at Selman for the duration of the war, except for brief interludes of advanced aircraft instrument study in Philadelphia, McClellan Field in Sacramento, California, and McDill Field in Tampa, where he worked on the new B-29 Superfortresses. Interestingly, while posted to Selman Field, because he was a local soldier, he was never required to live on base; he was able to go home after work every night. He was discharged from the Army in February 1946, at Camp Fannin, Texas.

After the war, Gates returned to United Gas, where he had a distinguished forty-three year career, retiring as district manager in 1972. He then worked with the Louisiana Utilities Commission until he retired from that job as general manager in1978. Afterwards he worked with Louisiana Gas Service as district representative until he retired for the last time.

Gates was married in 1941, a year before going on active duty. His wife Blanche was also actively engaged at Selman in field support. She opened the Post Exchange Cafeteria in 1943, and later opened the first civilian cafeteria. In 1944, she was placed in charge of personnel at the officers' mess and remained there until it closed in 1946. After the war, she began a long and successful career with the *Monroe News-Star* in newspaper advertising. Roy and Blanche had one daughter and three grandchildren. Blanche passed away in 1994, after fifty-three years of marriage.

Ellis Matson

Ellis Matson, of Albuquerque, New Mexico, entered the service at the age of twenty in 1943. After passing through a number of training programs, he was sent to Selman Field for navigation training. When he completed the course, he was assigned to the 5th Air Force, and made the long flight to Nadzab, New Guinea. There he was assigned to the 70th Troop Carrier Squadron. He was engaged in ferrying for the remainder of the war; he transported supplies and troops to combat areas and evacuated wounded troops, and in the later stages of the war, prisoners of war. As the war moved north towards the Japanese homeland, Matson's unit shifted to more forward positions: from New Guinea, to Clark Field in the Philipines, and then to Ie Shima, and finally to Japan itself.

Matson was discharged from the service as a captain in February 1946. He was awarded four battle stars: South Pacific, Leyte, Luzon, Ryuko, and Okinawa. He returned to Albuquerque, where he remained on active status in the reserves until 1957.

Eugene Klinedinst

Upon high school graduation, Eugene Klinedinst, of York, Pennsylvania, enlisted in the Army Air Corps. He passed through a number of military training facilities before entering Selman Field's advanced navigation training program. Following his graduation, he was assigned to the 433rd Troop Carrier Group in the South Pacific.

The mission of the 433rd was to drop paratroopers, move supplies, ferry troops to combat areas, and evacuate the wounded. Many of the flights were over enemy-occupied territory. Klinedinst's base of operations changed constantly with the fortunes of war. He served briefly in Australia, New Guinea, Biak Island, Luzon, Leyte, Iwo Jima, and Okinawa. His final assignment was occupation duty in Japan.

Following the war, Klinedinst remained in the Air Force and served in Vietnam in 1972. He retired in 1973, after thirty-one years of distinguished service. He then went back to school, graduating from the University of Pennsylvania's Wharton School of Finance. His awards included the Air Medal, Vietnam Service Medal, Combat Readiness Medal, Philippine Liberation Medal with two Battle Stars, Luzon and Leyte, the Asiatic Campaign Medal with Five Battle Stars, two Outstanding Unit Awards, Good Conduct Medal, and the Pennsylvania Commendation Medal.

Irving Leibowitz

Leibowitz, a native of Brooklyn, New York, arrived at Selman Field in 1943. The twenty-one-year-old was able to handle both the rigors of physical training as well as the classroom challenges and was commissioned a second lieutenant near the top of his class. Indeed, due to his achievements, he was ordered to remain on post at Selman Field and serve as an instructor of navigation.

In the spring of 1943, he met a local girl, Leslie Abrams, at a reception. After a year's courtship they married in 1944. Leibowitz lived off-base with his new wife and, except for periodic overnight navigation training flights with his students, he managed to be home each night.

With the downsizing of Selman Field and the end of the war, Leibowitz was sent back to New York City for discharge in 1945. He and his wife moved to Joliet and Chicago, Illinois, for a short time before returning to Monroe in 1947. They opened up a fabric shop, which proved successful, and they became actively involved in the civic and social life of the community. Shortly after the couple celebrated their fifty-sixth wedding anniversary, Leibowitz passed away in August 2000.

Paul Julian, Jr.

For as long as he could remember, Julian, wanted to be a fighter pilot. Julian, who was from Virginia, took the Army Air Corps cadet examination at Langley Field, Virginia, in 1942. After waiting for his results for several months, he finally received his orders in July 1943. He had already completed basic training and was sent to the University of Alabama for cadet training duty. Afterwards Julian went through pre-flight instruction in San Antonio and then passed through classification. His satisfaction upon discovering that he had qualified as a fighter pilot turned to disappointment when he was told the Army had an ample supply of pilots and needed navigators. Faced with the choice of going back into the army as a private or becoming a navigator, he chose the latter.

Julian was then sent to Selman Field, where he completed his navigator training. Following graduation, he returned home briefly on leave to marry his high school sweetheart, Nelma. Upon his return to Selman, he was ordered to Kearns, Utah, for overseas processing. Then, with brief stops for refueling, he made the long passage to Nadzab, New Guinea, where he reported to the 54th Troop Carrier Wing, 70th Squadron. The 54th moved north from New Guinea through the Philippines, Okinawa, and finally to the former Japanese naval station near Yokohama. The days were long and the missions not without problems. Troops and supplies were ferried at all hours of the day and night. In addition to the suffocating heat and high humidity which sapped their energy, Julian and his crew had to overcome fierce thunderstorms, typhoons, and primitive living conditions. They also had to cope with the threat of Japanese attacks. Nevertheless, they succeeded in accomplishing their missions. In January 1946, Julian rotated home. He remained on active flying status and stayed in the Air Force until he retired in 1957 with 8,600 military flying hours and over 100 hours as a civilian instructor.

HOWARD M. FISH

Howard Fish was born in the small town of Melrose, Minnesota, and raised in St. Cloud, Minnesota. He was educated in the city's parochial school system. After the Japanese attack on Pearl Harbor, Fish wanted to volunteer; however, because his brother, who was two years older, had already entered pilot training, his parents urged him to wait. He was so anxious to join the military that his parents ultimately relented. Fish signed up with the Army Air Corps in July 1942.

Fish was sent to the San Antonio classification center for evaluation and testing. After successfully passing all physical and psychological profile exams, he went into pilot training at Hatbox Field, Muskogee, Oklahoma. But after he accidentally sprained his ankle while playing basketball, he was put up for elimination due to his reduced mobility and what he called his "apprehension in the air;" although his proficiency was never in question, he was nevertheless bitterly disappointed. After that, he was sent to basic training at Wichita Falls, Texas. A call had gone out for volunteers for gunnery school. To escape the intensive physical regime, Fish was quick to volunteer; he was dispatched to Tyndall Field, Florida. After completing the course, he was retained as an instructor. The fact that many of his students were graduates of the pre-flight program at Selman Field led Fish to reassess his military goals. He decided to become a navigator and requested a change of orders to enter the advanced navigation training program at Selman.

Upon receiving approval, Fish arrived at Selman Field in February 1944. At about the same time, he received word that his brother, a B-24 bomber pilot flying out of Britain, was missing in action. The pain of this loss made him even more determined to finish his training and get into the war. Fish graduated at the age of twenty. He then joined up with his new crew, went through aerial training, and received orders to join the 301st Bomb Group, 15th Air Force, at Foggia, Italy.

The 15th Air Force was engaged in relentless raids against German targets throughout southeastern Europe. Fish remained unscathed on twenty-two missions, but on his twenty-third, on February 7, 1945, his bomber ran into intense anti-aircraft fire over Vienna, Austria. As the bombing run began, he had just logged the initial point when the bomber was hit by FLAK. The pilot punched the bail-out button and Fish jumped from the nose hatch. Instead of waiting as he had been taught, he immediately pulled the ripcord. His shoes, which were tied to the parachute harness, were ripped off, striking him in the face and breaking his nose and two teeth. He noticed almost immediately that his chute was ripped in at least six different places. Two of the holes were on either side of the seam; he feared that if the holes widened, the seam might not hold. He watched the holes continue to expand as he descended. Figuring he probably would not make it anyhow, he prayed, while continuing to eye his frayed chute. Suddenly, hearing a dog bark, he looked down and realized that his chute had held. He hit a muddy field hard and was dragged swiftly along the ground by a strong wind. He finally managed to punch the quick-release of the chute, which was blown downwind.

It was not long before a crowd of civilians and some soldiers had gathered around him. He was taken into custody, interrogated and sent to Stalag III, a prisoner of war camp housing captured Allied airmen located near Nuremberg, Germany. In April 1945, in an effort to escape the Allied advance, the Germans force-marched these prisoners to a new location. They were given little food or rest. On the second day of the march, with little warning, P-47s dipped low out of the clouds and strafed the column. Eight POWs were killed. That evening, the guards made a tactical decision to march the prisoners at night in order to avoid Allied air patrols. The exhausted and hungry prisoners trudged along in the rain and cold. After the guards had passed by, Fish and another prisoner decided to make a break for freedom. They fled into the woods, heading west toward American lines. Moving only at night and staying in the forest to evade detection, they advanced slowly due to their malnourished and weakened condition. As chance would have it, they were soon recaptured; however, less than a week later, they escaped again. This time they were more fortunate as they met up with forward elements of the 86th U.S. Infantry Division. Since the 86th was moving forward, they could not send Fish to the rear. So Fish stayed with these soldiers until Germany surrendered. Fish and his new comrades located a German car, painted U.S. markings on it, and Fish was finally able to drive back to his base in Italy.

Fish was flown home and given an extended 100 day temporary duty assignment at his home in Minnesota. At the time, former POWs were given their choice of station. Therefore, when his TDY was over, Fish requested transfer to Templehof Airbase in Berlin, where there was a vacancy for a control officer. His request was approved. Fish figured this was because the personnel officer felt that since navigators were good with numbers, they could learn statistical control. Fish joined a troop carrier unit airlifting men and supplies across the Soviet Zone of Germany to Berlin. All flights carrying a general required a navigator and Fish soon became the navigator for Gen. Lucius Clay, the commanding general of the theater.

Fish's comfortable postwar routine was soon disrupted by the Soviet blockade of Berlin, which in turn triggered the Berlin airlift. All flights in and out of Templehof were required to have navigators aboard to make certain the transport planes stayed within strictly authorized corridors. After advanced radio and radar navigation aids were introduced, the navigators became flight planners, briefing, and operations officers. Fish's work in statistical analysis attracted attention and he was transferred to airlift task force headquarters.

With the end of the blockade, Fish was assigned to the aircraft observer bombardment school at Mather Air Force Base near Sacramento, California, where he graduated third in his class. Due to his high marks, he was sent to Langley Air Force Base and became part of the Air Force's first jet bomber B-45 group. While at Langley, Fish applied for a regular Air Force commission. The squadron commander, a pilot, had a condescending view of navigators and refused to take action; he indicated that Fish, together with another navigator who had also applied, should wait until their next duty station to renew the request. The two navigators objected. A new squadron commander also gave the navigators negative endorsements on their applications and referred to the two men as agitators. The base commander finally settled the dispute, overruling the squadron commander.

In the meantime, the Korean War had broken out; Fish was initially deployed to Japan and later to South Korea. Fish flew sixty-four missions in Korea, mainly conducting night reconnaissance. Between missions, Fish just happened to be in a group singing around a piano at a makeshift bar, when the piano player, who was the personnel officer from 5th Air Force Headquarters, recognized him. He was looking for a navigator for Gen. Walton Walker and knew Fish from Germany. He selected Fish to be the navigator for General Walker, the Eighth Army commander.

After General Walker was killed, Fish served as the navigator for Gen. Matthew Ridgeway. The Chinese had entered the war and the Eighth Army was in full retreat. All flights were extremely dangerous. Nevertheless, Ridgeway insisted upon remaining by Fish's side in the nose of the B-17; he wanted to reconnoiter the front lines personally. Ridgeway aligned Fish's aerial charts with his artillery maps, trying to pinpoint enemy troop concentrations on the constantly changing battlefield.

One evening, the director of intelligence asked Fish if he would be willing to drop Allied agents into North Korea at night. Fish's extensive night reconnaissance flights with the general had convinced the intelligence chief that Fish was highly capable of carrying out the assignment. General Ridgeway gave Fish the go-ahead on condition that Fish volunteer. He was instructed to fly into North Korea, hugging the ground until he reached the precise drop zone to avoid alerting the enemy. Fish subsequently flew roughly half a dozen missions during which he dropped Korean male and female agents. They were equipped with large parachutes designed to open quickly so that the agents spent as little time as possible in the air and landed softly. The spies were not equipped with jump boots or any other equipment; they were dressed in typical Korean peasant clothing so that they could blend in with the populace. They were provided with short-range radio transmitters to relay reports at pre-arranged times and places. Fish later reported that these men and women performed heroically.

After President Truman fired General MacArthur, Ridgeway took his place. Fish returned to Langley Air Force Base in Virginia as squadron navigator of a B-26 combat training outfit. With the end of the Korean War, his group was selected to be the first combat unit to receive the new British-built B-57 Canberra bomber. Following a brief period of study in Britain in order to familiarize himself with the Canberra, Fish returned to Langley as the navigator of the new B-57 group. Within a short time, he became the director of operations at Langley. However, the new commander was not pleased with a navigator as his director of operations. Realizing that for further advancement he would need additional training, Fish resolved the problem by volunteering to go to the air command and staff college. Fish was accepted and graduated number one in a class of 520. He then returned to Langley as the wing executive officer.

In the years that followed, Fish furthered his education at the armed forces staff college and the air war college. He earned a Masters of Business Administration at the University of Chicago and a Masters of International Affairs at George Washington University. He then served four years as the chief of analysis for air force plans in the Pentagon.

Fish knew that he was due for reassignment. With the Vietnam conflict underway, he volunteered for duty in Vietnam. Gen. George Brown, the Air Force Commander in Vietnam, was seeking as director of analysis a man with strong analytical skills. The Air Force chief of plans and operations recommended Fish for the position and the job was his. Fish created an outstanding team of fifty professionals for General Brown, who subsequently became chairman of the joint chiefs of staff. For his outstanding work and many years of service, Fish was promoted to brigadier general, the first graduate of Selman Field to attain such a high rank.

When he left Vietnam, Fish went on to serve in a number of distinguished positions, including the director of the air force budget office, deputy assistant secretary of defense for international security affairs, chairman of the U.S. military delegation to the United Nations, and assistant vice chief of staff for the United States Air Force. After thirty-seven years of service, Fish retired on June 1, 1979. Since 2004, he has lived with his wife of fifty-nine years in Shreveport, Louisiana, where his son, an assistant district attorney, also resides.

General Fish modestly credits what he learned at Selman Field for much of his success: "I think my early training at Selman Field gave me unique opportunities. I've had a great career and the foundation was laid at Selman."

Robert Wilson

Bob Wilson was born in Sacramento, California. When he was very young his parents moved to Peoria, Illinois, where he attended the local public schools. After graduation from high school, he entered the University of Illinois to study mechanical engineering. The war was already raging and Wilson knew it would only be a matter of time before he was drafted. After completing his first semester, he dropped out of university and volunteered for the Army Air Corps. Following induction in early 1943, Wilson was assigned to pilot training in Tuscaloosa, Alabama. However, a lack of depth perception disqualified him for that program, and he was asked if he wished to go to navigation school. He agreed and was sent to Selman Field for training.

Wilson found the coursework challenging. The strict schedule of classes and exertion of the physical exercises were exhausting. Indeed, he found himself always in need of a little more sleep. Moreover, he never adjusted to the numerous insects that infested Selman Field. He was amazed by one of the other cadets in his barrack, a Chinese student by the name of Willie Sing, who considered the insects an asset rather than a liability. Sing would capture the black widow spiders found under the wooden walkways that served as sidewalks. He would then send them to his parents in Camden, New Jersey; they were in a business that utilized spider webs for certain optical equipment needed by the military.

Wilson completed his training at Selman Field in December 1943, just shy of his nineteenth birthday. He was then assigned to a B-17 bomber crew. He and his crewmates spent several weeks going through simulated combat training in Avon Park, Florida. Then they were ordered to Langley Field, Virginia, where they endured a rigorous radar/bombardier air/ground training program. Following the successful completion of the course, the crew was ordered to fly to Foggia, Italy, to join the 15th Air Force. Taking the circuitous southern route, the B-17 made fueling stops in Puerto Rico, Brazil, North Africa, and finally Italy.

Upon arrival in Foggia, Wilson was separated from his crew and assigned to perform radar work on a specially equipped plane that would serve as the lead bomber on future raids. The ball turret had been removed and replaced with a radar antenna that Wilson monitored from the cramped radio room. The intent was to improve the efficiency of the bombing runs over targets hidden by fog or cloud cover. The Germans had grown adept at using smoke pots which obscured important targets on visual bombing runs and the 15th Air Force believed that this new device could thwart the German's camouflage tactics.

The purpose of the strategy was to allow the lead crew to adopt a course of action that would be followed by others. If the target was difficult to see, Wilson would toggle the bombs and all the planes behind his in tight formation would release their bombs simultaneously. It was hoped the Germans would be caught off-guard and surprised by the pinpoint bombing through the clouds. Initially only a half dozen planes in the 15th Air Force were configured for radar bombing. All of the navigator/bombardiers lived under constant stress. Air losses were heavy and the only way to get through each run was for the men to believe that no matter what happened they could manage to survive.

Each flight demanded hours of careful study. On the day before each mission, Wilson would be flown to Air Force headquarters to study the radar image of the next day's target; he would memorize every detail. The 15th Air Force was conducting a relentless and massive campaign to destroy the oil refineries at Ploesti, Rumania; these vital targets were well known, as their destruction could cripple Hitler's ability to wage war. Early in the morning of July 15, 1944, Wilson set off on his third raid in the lead B-17 bomber against Ploesti. FLAK was heavy over the target. Suddenly a burst of metal tore through the outer skin of the plane. Two engines failed and gasoline began to gush through the fuselage. Norton, the engineer, was knocked unconscious with a piece of FLAK embedded in his helmet. Wilson was able to remove it with a pair of pliers and Norton came to just in time to hear the pilot order everything not fastened down to be thrown out in order to lighten the aircraft. Guns, ammunition and everything else that was loose were tossed out. The crippled plane slowed, but remained airborne; the rest of the squadron formed a cordon around the plane to protect it against German fighters.

After two hours, the third engine started smoking. The pilot indicated that they were over Serbia, somewhere south of Belgrade. He said he could no longer control the plane and ordered everyone to bail out. He and the co-pilot stayed with the plane a little longer to steady it for the forced evacuation. Although

Wilson never saw them again, he did learn that they had been able to bail out safely before the plane crashed.

Wilson was relieved to be out of the burning aircraft. As he approached the ground, he could see the farmers in a nearby field pointing up at him and running towards him. Before he could flee, an older man approached him and asked "Amerikanski or Englace." When Wilson replied "Amerikanski," the old man hugged him, and kissed him on the cheek. He then motioned to a young girl carrying a wooden keg to come forward. He took the keg and offered it to Wilson who was very thirsty. Wilson drank deeply and came up sputtering. It was not water, but Slivovitz, a very powerful homemade plum brandy. After they all had a good laugh, Wilson was guided to a Chetnik military field office, where his crew was slowly assembled. Seven of the eight crewmen had landed safely. Only the radio operator suffered an injury, having broken his ankle in a rough landing.

Wilson and the other airmen found the Serbs to be hospitable. The men felt less anxious and were thankful they had fallen into friendly hands. Through sign language and broken English, they learned that they had parachuted into territory controlled by Gen. Draza Mihailovic's Chetnik fighters. The Chetniks were fighting a two-front war simultaneously: against Tito's Communist partisans as well as against the Germans.

The Chetniks provided the airmen with two guides wearing Yugoslavian military uniforms, and an English-speaking interpreter who had previously lived and worked in the steel mills of Gary, Indiana. The men spent the better part of the next two months walking across the country toward Pranjane, where General Mahailovic was quietly assembling over 500 downed Allied airmen in hopes of a rescue. The men were kept constantly on the move in the mountains to avoid the Germans. At night, they stayed with peasant families who shared with them what little food they had.

The general had major problems in planning the rescue. The British sided with Tito and therefore, provided no diplomatic or military support to Mihailovic. The United States had no presence in the country and no communication with the Serbs. The airmen already in Pranjane decided to try to assemble a radio from the parts scavenged from a number of crashed planes. They managed to accomplish this, but the 15th Air Force Headquarters in Italy had no idea of the huge numbers of downed fliers that Mihailovic was protecting, and initially thought their broadcasts for help were bogus. The airmen finally developed a code that they hoped might get past the Germans, yet be understood at their base.

The code worked and the OSS parachuted three operatives in to organize the rescue. The operation was termed the "Halyard Mission." The airmen and the Serbs built a makeshift airfield in a pasture on the mountainside, and on August 9, 1944, the first four twin-engine C-47 transports began what would be the largest single rescue in the history of World War II. Over the next few weeks, more than 500 Allied airmen were recovered, without the loss of a single man. Several thousand Chetnik fighters protected the airfield as the men were airlifted out. Wilson boarded one of the last flights.

Wilson was sent home to Langley Field, Virginia, where he headed up the radar school. He was discharged in September 1945, after the Japanese surrender. He resumed his studies at the University of Illinois, completing his degree in mechanical engineering. He was employed as an engine designer and analyst for Caterpillar Tractor Company for eighteen years. In 1967, he married his wife, Sharon; they still reside in Peoria.

After the war, Yugoslavia fell under Marshall Tito's control. He put General Mahailovic on trial for treason, claiming that he had collaborated with the Germans. In fact, Mahailovic's guerilla tactics had kept the Germans tied down and forced them to postpone their invasion of Russia. This later proved to be a major factor in their failure there. Wilson and a number of other rescued airmen formed "The Committee to Rescue Mahailovic," hoping to testify on his behalf at the trial; unfortunately, they were not permitted to do so. Tito had Mahailovic executed and his remains placed in an unmarked grave. Thereafter, schoolchildren in Serbia were taught that he was a traitor.

Pres. Harry Truman presented the Legion of Merit as commander in chief of Yugoslavia to the general posthumously for his efforts on the airmen's behalf and for his overall effort in the war. The award could not be delivered and was placed in the State Department archives until 2005.

In 1966, Wilson vacationed in Tito's Serbia and tried to locate some of his rescuers to thank them personally. He did manage to locate a few of them in the tiny village of Bunar. The people remembered Wilson falling out of the sky and held a huge celebration, killing a suckling pig at a festive party in his honor. Several years later, in 2004, he made another trip to Serbia. Wilson, his wife, as well as another rescued airman, Clare Musgrove, his grandson, and two OSS officers, George Vujnovich, OSS director of the Halyard Mission, and Arthur Jibilian, the radio operator who had parachuted in for the mission, and his wife were guests of the foreign minister. At the dedication of a monument at the airfield, Wilson gave a moving talk about their rescue, thanking the Serbs for their efforts.

In 2005, Wilson received yet another invitation from the foreign minister, Vuk Draskovic, and the American embassy in Serbia to once again be their guest at the awarding of the Legion of Merit to Dr. Gordana Mihailovic, a retired M.D. and the daughter of the general. The other invitees to the ceremony included those who had been with Wilson on the 2004 trip. Dr. Mahailovic, after receiving the award, presented the medal to the Serbian Orthodox Church Museum so that the public would be aware that her father truly was a national hero.

John H. McCarter, Jr.

John H. McCarter, Jr. was born in 1921 in Utica, Mississippi, just south of Jackson; he was educated in the Jackson public school system. After he graduated in 1939, he enrolled at Hinds County Junior College in Raymond, Mississippi. Upon completing the two-year program, he went to San Diego, California, and worked at the Consolidated Aircraft Company producing B-24 bombers. It was at that time that he was drafted and sent to Camp Shelby, Mississippi, near Hattiesburg, for basic training; he completed training at Keesler Field in Biloxi, Mississippi.

From Biloxi, McCarter was sent to Ypsilanti, Michigan, to attend a special school on aircraft maintenance; he finished with some of the highest scores in his class. While at Ypsilanti, he was elated to be introduced to Henry Ford and Charles Lindberg.

After Ypsilanti, McCarter was assigned to gunnery school in Laredo, Texas. When he finished this training, he was sent to Boise, Idaho, as a turret gunner. He met up there with his crewmates and began training in a new B-24. He also came into contact with Hollywood legend, Jimmie Stewart, who was one of the commanding officers on base. When simulated combat training was completed, the crew was assigned to the South Pacific and began the long flight to an airbase south of Darwin on Australia's north coast. From there, the airmen flew bombing missions against Japanese-held islands throughout the southwest Pacific. The crew emerged from their tour of duty unscathed and McCarter returned home; he was assigned to Miami Beach for some well-deserved R & R.

When McCarter completed his furlough, he was sent to Chanute Field, near Champaign, Illinois. There he was finally hospitalized with malaria, although he had had symptoms since his return from the south Pacific. In the late spring of 1945, McCarter was discharged from the base hospital and assigned to Selman Field as a technical sergeant in charge of "head of the line." His duties were to conduct daily inspections of all planes on the flight line, as well as pre-flight inspections of all planes, and to make any maintenance adjustments that were necessary. During his time at Selman Field in the fall of 1945, until he was discharged, no planes were lost on his watch.

While McCarter was stationed at Selman, all personnel were required to practice on the rifle range with the M1 carbine rifles. McCarter scored 189 out of 200 on the shooting tests, the highest score on base, and received an "expert rifleman" award.

With the war in Europe over and the conflict with Japan nearing its grisly end, there was less stress among the men and there were even some humorous moments. After heavy rains, McCarter noticed that many skunks which lived under the barracks would be flooded out and would run in the open on the base. The men had to take wide detours to avoid causing the animals to release their pungent odors.

Towards the end of the war, German prisoners of war were used to pick pecans off the trees in the adjoining Stubbs Pecan Orchard and unload cargo destined for Selman Field off the train on the nearby railroad spur. Many also performed routine custodial and maintenance jobs on the base itself. One POW, a carpenter, was assigned to fix a lock on McCarter's barrack. When McCarter returned to the barrack, he found the German had satisfactorily completed the job and then had taken a nap on his bunk. Both men got a good laugh.

McCarter was discharged at Ellington Field, near Houston, Texas, in 1945; he went on to LSU on the G.I. Bill to pursue a B.S., M.S., and Ph.D. in geology. While still enrolled in school, he worked for the state Conservation Department. He also married at this time. After completing all his coursework for his Ph.D., he left LSU and taught geography at Louisiana Tech for several years. In 1951, he was asked to start up the geography department at nearby Northeast Louisiana University, now the University of Louisiana at Monroe. He served as department head for ten years, until 1961, and then went into private consulting in the oil and gas business in northeast Louisiana. However, he was asked to return to teaching on eight separate occasions at Northeast Louisiana to help maintain the program's excellence, and did so, teaching a variety of geography courses.

McCarter and his wife, Louise, had three children. Louise passed away in March 2005, and her demise was followed by the death of their youngest son two months later. McCarter has remarried and continues to reside in Monroe.

Theodore "Dutch" Van Kirk

Of all the Selman Field navigators, perhaps the best known is Theodore "Dutch" Van Kirk. He earned his place as a legend in history by serving as the navigator aboard the "Enola Gay" which dropped the atomic bomb on Hiroshima.

"Dutch" Van Kirk was born in 1921 in Northumberland, Pennsylvania. When he was nineteen and in his first year of college, he joined the Army Air Corps, as he believed that the United States would soon be at war. It was not until the following year, 1941, that he was finally inducted. Following basic training, he was sent to Kelly Field, Texas, for navigation training, graduating as a second lieutenant in April 1942. He was assigned to the 340th Bomb Squadron, 97th Bomb Group, the first operational B-17 Flying Fortress unit in England.

Van Kirk flew on the "Red Gremlin," along with pilot Paul Tibbets and bombardier Tom Ferebee, two men who would later fly with him on the histotoric Hiroshima mission. This crew served as the lead aircraft in the 340th Bomb Squadron, carrying out attacks on targets throughout occupied Europe. From August 1942, until June 1943, Van Kirk flew fifty-eight combat missions and eight transport missions to Europe and North Africa. Several of his transport missions were enshrouded in secrecy and involved meticulous navigation over long distances and over water.

In October 1942, Van Kirk and his crew took part in the first of two all-important missions. First, prior to the invasion of North Africa, they clandestinely flew Gen. Mark Clark from England to Gibraltar for secret negotiations with the Free French. Then, the following month, the crew once again made the same 1,000-mile trip over hostile waters to Gibralter, this time carrying General Eisenhower and members of his staff. Upon arrival, Eisenhower assumed command of the invasion forces for "Operation Torch."

After delivering Eisenhower, Van Kirk and his crew were transferred to North Africa; from there, they participated in bombing runs on German installations. On one mission, the crew fueled their bomber and loaded their own bombs for a raid on Bizerta, Tunisia. At the time of the raid, the crew did not have any idea whether it had been successful. It was only after the war that Tibbets learned that the bombs dropped on the air base had hit the mess hall where some of the most experienced German pilots were eating, thus severely crippling base operations. "Just goes to show, you don't have to be good, just lucky sometimes," Van Kirk later recalled.

In June 1943, Van Kirk was ordered to return to the United States and report to Selman Field as an instructor. At Selman, he took issue with the way the navigators were being trained: "They trained the navigators at 7,000 feet. Things look a lot different at 20,000 feet." He remained at Selman Field until October 1944, at which time he was transferred to the 509th Composite Wing, based at Wendover, Utah. Here, Van Kirk was reunited with his former crew. Tibbets, now a colonel, had selected Van Kirk to be the group navigator. At the age of twenty-four, Van Kirk was the "old man" of the crew.

From November 1944, until June 1945, the crew underwent intense and secret training in the new B-29 Superfortress in preparation for dropping the atomic bomb. On June 25, 1945, Tibbets and his crew arrived on the island of Tinian, where they continued their clandestine training under heavy military guard. On July 16, they learned of the successful test of the atomic bomb in the New Mexico desert. Detailed briefings and information sessions concerning the bomb and its aftershocks followed. After President Truman issued the order to drop the bomb, the countdown began. Training accelerated and grew more intense. The target, Hiroshima, was the Japanese army headquarters for defense of the home islands. It housed over one hundred significant military targets. The crew learned that, following the drop, they had to leave the target area as quickly as possible. "The scientists told us we had to be eight miles away from the blast for the plane to survive," Van Kirk later recalled, referring to the warning about releasing the bomb; "but they really weren't that sure. We had stripped down the airplane so we could get high and get away fast after we dropped it."

By August 4, 1945, all preparations for the mission had been completed. The skies over Japan were clear of cloud-cover. Tibbets had the words "Enola Gay," his mother's maiden name, painted in bold black capital letters under the pilot's window on the port side of the B-29. The following day, August 5, the atomic bomb, "Little Boy," was wheeled out of its assembly hut amid massive security; it was then carefully loaded into the modified bomb bay. The uranium bomb was huge, ten feet in length; it weighed 9,700 pounds.

That evening the seven crews going on the mission gathered for the first of two late night briefings: three weather planes; two aircraft, one of which carried cameras, the other, special instruments to record the blast; the crew of the "Enola Gay," and a seventh plane that was to fly as far as Iwo Jima and then stand by. At the final meeting, held at 11:00 P.M., only the crews of the "Enola Gay" and the two planes loaded with equipment that would accompany her on the mission were present. Here, for the first time, the crews heard about the massive destructive power of the bomb. When the men learned that the bomb had the power of 200,000 200-pound bombs, they sat in stunned silence. The briefing officer continued, emphasizing escape maneuvers to be employed immediately after the bomb was released, procedures the crews had been practicing during their training.

When the meeting broke up, the men were told to get some sleep. This proved impossible for Van Kirk, as he weighed the enormity of the mission. Shortly after 1:00 A.M. on August 5, the crews were taken by jeep to the flight line; there they saw the planes lit up by spotlights. Take off was scheduled for 2:45 A.M. There was one last pre-flight check of the "Enola Gay." It had been stripped of all non-essentials to make it 6,000 pounds lighter in order to compensate for the weight of the bomb. Flight time for the 1,700-mile round trip was estimated to be twelve hours and fifteen minutes. Van Kirk remembered the mission as "Rather routine... and one of the reasons for it was it went exactly according to plan." As the plane set out, Van Kirk carefully plotted the course-setting over the ocean. Fifty miles out, the men could see Hiroshima clearly. Cruising at 31,000 feet, there was no enemy fighter opposition. Van Kirk's reckonings took the plane over the target within seventeen seconds of the scheduled time. When the bomb was released, the plane's nose jerked upward, and then Tibbets put the bomber into a sharp 180-degree turn. Van Kirk gave Tibbets a new heading back to Tinian. Forty-three seconds after the bomb was dropped, with the plane at the predetermined altitude of exactly 1,890 feet above the ground for maximum impact, the bomb detonated. The sky lit up brighter than the sun. In spite of their dark goggles, the crew felt as if someone had shone a bright light in their eyes. Before long the plane was eleven miles away from the target; nevertheless, shock waves continued to make the plane tremble. An enormous mushroom-shaped cloud arose nine miles from ground zero, three miles above their cruising altitude. Van Kirk could see some of the effects of the explosion: the ground was covered with back smoke and dust. He was just thankful he and the crew had been spared. As the "Enola Gay" headed home, Van Kirk and the others hoped the mission would bring an end to the war.

After the war, Van Kirk participated in "Operation Crossroads," the first Bikini Atoll atomic bomb tests. He left the service as a major in August 1946, having received the Silver Star, the Distinguished Flying Cross, and fifteen Air Medals. Upon reentering civilian life, he completed college, receiving his B.S. and M.S. degrees in chemical engineering from Bucknell University. After graduation in 1950, Van Kirk spent the next thirty-five years working for Dupont in technical and managerial positions in research and marketing. He was also a much sought-after speaker at veterans' meetings all over the country.

At the twelfth Selman Field reunion in Monroe in April 2006, Van Kirk was the keynote speaker. He was given a special medal from the Selman Field Historical Association, honoring him for "sixty years of practicing with a pistol." The award, given in jest, referred to the requirement at Selman Field that airmen had to qualify with a .45-caliber pistol. According to Van Kirk, "The only person on the base with a lower score than me was the base urologist."

Endnotes

[1] German acronym for *fliegerabwehr kanone,* a flyer defense cannon; better known as anti-aircraft fire.

Epilogue

Perhaps the best way to conclude this tribute to the brave men who served at Selman Field would be to recall the words found on a Selman Field web site:

To all the men who passed through the gates of Selman Field, however briefly, and for whatever reason; we honor you today and always. Our appreciation for what you did during the time of the second world war of the twentieth century will not be forgotten.

~author unknown

Bibliography

Primary Sources
Interviews

Album, Rose Marie, Monroe, LA, September 17, 2004.

Alston, Maurice, Monroe, LA, April 30, 2005; April 29, 2006.

Armstrong, Colonel George, Monroe, LA, February 23, 2007.

Bailey, R. Burrell, Monroe, LA, April 30, 2005.

Barham, Sen. Robert, Monroe, LA, September 28, 2006.

Blumenthal, Morris, Monroe, LA, September 22, 2004.

Blumenthal, Rita, Monroe, LA, September 22, 2004.

Bowie, Georgette, Monroe, LA, April 30, 2005.

Fish, Howard, Monroe, LA, June 12, 2007.

Gates, Roy, Monroe, LA, April 29, 2006.

Humphrey, Sam, Monroe, LA, April 30, 2005; April 29, 2006.

Kelly, Ethel, Monroe, LA, April 30, 2005; April 29, 2006.

Leibowitz, Leslie, Monroe, LA, April 30, 2006.

McCarter, John H., Monroe, LA, June 24, 2006.

Marcovitz, Jean, Monroe, LA, September 22, 2004.

Mintz, Jean, Monroe, LA, September 17, 2004.

Parsons, Carol, Monroe, LA, October 3, 2006.

Peterson, Harry, Monroe, LA, April 30, 2005.

Rorke, George, Monroe, LA, April 30, 2005; April 29, 2006.

Strauss, Ernest, Monroe, LA, January 31, 2005.

Weems, Don, Monroe, LA, September 23, 2004.

Wilson, Bob, Monroe, LA, May 27, 2007.

Manuscripts

Homdrom, Theodore, personal memoirs.

Joiner, Billie, and Berry Jean. "The History of Aviation at Northeast Louisiana University" (unpublished paper), 1985.

McInosh, J.F., personal memoirs.

Rifkin, Joseph, personal memoirs.

Shankin, Earl, and Courtney, personal memoirs.

Siantz, Edward, personal memoirs.

Titkemeyer, Charles, personal memoirs.

Newspapers

Monroe Morning World, March 1942-September 1946.

Selman Field Beacon, 1992-2006.

Legal Documents

Ouachita Parish Police Jury. Resolution by Police Jury, Meeting of October 17, 1941. Ouachita Parish Court House, Monroe, LA.

United States vs. Ouachita Parish Police Jury, et al. June 17, 1942. Conveyance Book 297, 711. Conveyance Records of the Ouachita Parish Clerk of Court, Monroe, LA.

United States to City of Monroe. September 8, 1949. Conveyance Book 446, 431. Conveyance Records of the Ouachita Parish Clerk of Court, Monroe, LA.

Secondary Sources
Books, Typescripts, and Theses

Morse, Richard E. *Personal Papers in the United States Air Force Historical Research Center: Personal Papers of Earl Naiden, 1911-1944.* 5th ed. Maxwell Air Force Base, AL: US Air Force Historical Research Center, 1990.

Wings Over America: United States Army Air Forces. Baton Rouge, LA: Army and Navy Publishing Company of Louisiana, 1942.

Ambrose, Stephen E. *The Wild Blue.* New York: Simon and Schuster, 2001.

Baylis, Charles D., ed. *Yearbook of Selman Field: Monroe, Louisiana, 1943.* Baton Rouge: Army and Navy Publishing Co., 1943.

Hallack, Robert. *An Army Air Forces Navigation School.* Special Collections. University of Louisiana at Monroe.

Harbold, Norris B. *The Log of Air Navigation.* San Antonio, TX: The Naylor Co., 1970.

History of the 80th Flying Training Wing. USAF Historical Division Archives, Maxwell Air Force Base, AL.

Jones, Randolph. "Otto Passman and Foreign Aid." M.A. thesis, University of Louisiana at Monroe, 1982.

Selman Field: World War II: 1942-1946. Paducah, KY: Turner Publishing Co., 1998.

Selman Field Base Operations. "History of Base Operations and Affiliated Departments." July, 1942-February, 1945. Copy obtained from Swansey Kotz.

"Selman Field Log: 1942-1945." Copy obtained from Roy Gates.

The Official Guide to the Army Air Forces. New York: Simon and Schuster, 1944.

Wright, Monte D. *Most Probable Position: A History of Aerial Navigation to 1941.* Lawrence, KS: University of Kansas Press, 1971.

Index

Chennault Aviation and Military Museum

The museum is located in one of the few buildings remaining on the old Selman Field Army Air Corps Navigation training base. The museum has as its mission statement: "To preserve and promote military and aviation history and the legacy of the men and women who contributed to it."

The feel of a still-sturdy uniform, the gleam of lovingly polished medals, the curve of a well-used spoon, and the crooked lip of a bent metal cup—history is told through the stories of ordinary things and extraordinary men and women. Louisiana plays a pivotal role in our nation's aviation and military history, which the museum strives to promote and preserve.

The museum opened in 2000 with a small cadre of volunteers and is now part of the Louisiana secretary of state's museums division. It is housed in one of the last remaining buildings used by the Selman Field Army Air Corps Navigation School. The museum is also supported by a non-profit organization of Friends who are committed to providing funding, volunteers, and other resources.

The museum features a variety of historical exhibits about Louisiana natives Gen. Claire Lee Chennault and astronaut Jim Halsell, as well as exhibits on Selman Field itself. Additional exhibits include vintage aircraft and radio equipment, space travel, and the history of Delta Air Lines. Many of the displays feature individual servicemen or women and their contribution to our country's freedom.

History permeates the building, from the original heart of pine floors and wainscoting on the walls to the stories of individuals who gave of themselves to protect the precious freedom we all enjoy. There are stories of some who walked away from an airplane crash, a battlefield, a prison camp, and of some who did not survive. The stories are inherently different—some are sad, some are comical, yet the central thread of each remains the same—they are all about dedicated individuals caring enough for an ideal that they risked personal safety to protect it.

Completed in 2007, a major renovation allows use of the entire 9,200-square-foot building for exhibits and administrative use. Future plans include building a large hangar dedicated to Chennault, which will house vintage aircraft as well as displays.

The 2008 session of the Louisiana legislature approved a bill changing the name of the museum to the Chennault Aviation and Military Museum in an effort to bring international attention to the museum. All existing displays relating the story of Selman Field will remain in place.

The Selman Field Historical Association

The desire of Selman veterans to reunite goes back to the early 1970s. It finally materialized in 1986 when the convention and visitors bureau sponsored the first Selman Field reunion. This gathering was an overwhelming success, and the veterans voiced their desire to continue these reunions. A steering committee was formed to plan and execute the second reunion in 1989. After two well-recieved receptions, the veterans suggested forming an association to perpetuate biannual reunions. In 2005, it was decided that the reunions become an annual event.

Our membership is composed of both military and civilian men and women who trained, worked on, or were stationed at Selman Field, family members of those who were at Selman, and individuals who wish to support the goals of the association. Our membership is both national and international in scope. The official newspaper of the association is the *Selman Field Beacon,* which members receive three times a year.

The Selman Field Historical Association has the distinction of being the oldest historical association in northeast Louisiana that focuses exclusively on the Army Air Forces Navigation School at Selman Field during World War II.

Biannually since 1989 and annually since 2005, there has been a Selman Field Historical Association Reunion. Veterans, their wives and families from across the nation and even over seas, have attended, met old friends and classmates, and made new friends before returning home. Making these reunions the very best they could be has always been the main goal of the Selman Field Historical Association. The feedback from the veterans is overwhelmingly favorable, which lets us know that we have met our goal. This success would not have been possible without the support of the community.

This generation of Americans possesses a sense of history that is unparalleled by others. They have been hailed as the greatest generation of Americans and rightly so. By honoring the Selman veterans, we are not only acknowledging the enormous debt owed to them, we are passing the torch of knowledge of the people and events of World War II to future generations.

Selman Field Historical Association Mission Statement

The Selman Field Historical Association was founded with three goals:

I. To provide reunions for former Selman Field World War II Army Air Forces Navigation School personnel, graduates, and their families.

II. To collect and care for memorabilia from the World War II era with particular emphasis on aerial navigation and Selman Field.

III. To establish and maintain various memorabilia displays on aerial navigation and Selman Field and to provide speakers for schools and civic organizations interested in the history of Selman Field and aerial navigation.

IV. To serve as an archive of current holdings, donations, and ongoing research. The archive will be a repository of expanding data on Selman Field, its people, and aerial navigation, which will be available to researchers upon request.

The Selman Field Historical Association filed Articles of Incorporation with the state of Louisiana as a non-profit corporation, 501(c)3, on September 12, 1989.